I Like to Feed
DINOSAURS!

Jeff Ashley

ISBN 978-1-68517-856-7 (paperback)
ISBN 978-1-68517-857-4 (digital)

Copyright © 2022 by Jeff Ashley

All rights reserved. No part of this publication may be reproduced, distributed, or transmitted in any form or by any means, including photocopying, recording, or other electronic or mechanical methods without the prior written permission of the publisher. For permission requests, solicit the publisher via the address below.

Christian Faith Publishing
832 Park Avenue
Meadville, PA 16335
www.christianfaithpublishing.com

Printed in the United States of America

I like to feed dinosaurs! Feed time is in the afternoon every day. If I don't feed them every day, they could freeze to death at night. It only cost fifteen dollars a month to feed them if you buy dinosaur feed from Tractor Supply. I started this hobby as a way to honor my mother who passed away ten years ago. She fed dinosaurs too. You see, dinosaurs evolved into birds during the Jurassic time period 150 million years ago. The Museum of Natural History states that "today, we can safely declare that birds evolved from a group of dinosaurs known as maniraptoran theropods—generally small meat-eating dinosaurs that include velociraptor of Jurassic Park fame."[1] Therefore, my birds are the great, great, and lots of greats offspring of dinosaurs. Pretty cool!

    I grew up in Westchester County, New York, in a nice rural neighborhood with lots of trees and woods that encroached our home. I bought a chain saw when I was sixteen, and one day, my father gave the task of thinning out the woods behind our home. Everything that was less than two inches in diameter was fair game. This kind of job suited my type-A personality to a T. Within two hours, I was done. I could see a mile in every direction. I quantified my work as PDG—pretty darn good. Well, that made my mom PDM. That's pretty darn mad. I inadvertently cut her favorite bird tree perch. It was no big deal to me at the time. Today, I can see her point. A good perching tree for incoming birds to a bird feeder is as important as having a runway at the airport. Very essential. I have four bird feeders in the back of my garage office—man cave. Sitting at my desk, I enjoy watching the chickadees fly up to the bird feeder, take one seed at a time, and fly to a nearby small bush/tree thing. My wife said it would really get big when she planted this six-inch sickly-looking seedling. Today it is a Goliath bush that birds love to use as a runway/perch to the bird feeders. Chickadees are my favorite type of bird. They always seem to be polite and patient. They would land on the Goliath bush and wait for their turn to land at the bird feeder one at a time. They would take one sunflower seed, fly back to a perch, hold that seed between their quarter inch feet, open it up, and eat the seed. How many seeds in an hour does a chickadee eat? It should be quite a lot being that they were once dinosaurs.

If you feed birds a 150 million years ago, it would have eaten you, and I find it a bit paradoxical. Nonetheless, if you feed birds for over ten years, you create an interesting microenvironment that brings a plethora of wildlife. Chickadees are the predominate bird in my Adirondack north country home, then the bully blue jays, nuthatches, wrens, an occasional cardinal, squirrels (six today), a rare redheaded woodpecker, and now turkeys. So many birds came to my original one bird feeder that the runway became all clogged up. The polite little chickadees stared, getting nasty with each other, not able to fill their dinosaurian appetite, so I added three more over the years. Two of largest feeders are Audubon deluxe squirrel-proof bird feeders that hold about ten large plastic McDonald's soda cups full of black sunflower seeds. My wife said I was being mean to the squirrels, so I de-squirrel proofed them. The squirrels hang upside down, eating tons of sunflower seeds and dropping just as much as they eat. This helps feed the wild turkeys. Yesterday we had a heard of fourteen. Some would say a flock, but from an evolutionary standpoint, I say a herd. Now in the Adirondacks of Northern New York, where I live, turkeys have made a major come back in population size. These are amazing birds. Their heads can change color from excitement or emotion to red, white, pink, or blue. This is true. When there were fourteen turkeys in my backyard yesterday, I saw one with a blue head and thought to myself, *That's kind of weird.* A blue head indicates the bird is stressed. I heard that they roost in trees, and I had never witnessed that until two days ago when one of my daughters who came home from college and I went for a walk in the woods behind our garage. This is the Adirondacks. When I say woods, you might think of a few acres, but in my backyard, you can walk several miles before the woods ends. When you startle a pheasant in the woods, it takes off with a loud *whoosh, whoosh* that usually scares you half to death. Well, a turkey weighs about as much as 15 pheasants and sounds like a condor crashing through the woods when you disturb it. Turkeys are not graceful in flight. So when my daughter and I startled one on our walk near sunset, it was roosting about forty feet up in a large pine tree. Its wings clipped numerous branches as it took off. A few seconds later, we started two more. That's why they don't like to fly; they just stink at flying. My chickadees, on the other hand, are beautiful in flight. They can fly through a tangle of branches and never touch one. Perhaps turkeys evolved ten million years later than chickadees. Being rotten flyers, turkeys are always walking. It is said that they can run up to twenty-five miles per hour and fly at up to fifty-five miles per hour, but not through a tangle of branches.

It is wintertime in my hometown of Wilmington, New York. In the Adirondacks, it is common to have a continuous snow cover on the ground from December through part of April. We as humans tend to wear boots in the winter when there is snow on the ground. But at some point, in the life of an Adirondacker, we have all walked on the snow barefoot for a minute or two for some dumb reason like running to the car to get a forgotten cell phone. We only do this once in a lifetime as there is an imprinting on the brain of the feeling you get when walking barefoot in the snow. Turkeys do this for five months of the year. How is that possible? I was thinking about that question one day as I watched the turkeys under my bird feeders. What I found out is that turkeys have what is called countercurrent blood circulation in their feet and legs where the veins and arteries in their legs are so close and intertwined with each other, they act as a heat exchanger. The warm blood flowing from the heart is cooled down before it reaches the feet, and the cold blood circulating from the feet is warmed up before reaching the heart. The blood that reaches the feet is just barely warm enough to keep the turkeys' feet from freezing, and because the cold blood returning to the heart is warmed up, it prevents the turkey's body from cooling off too much. It is a miraculous way of preventing frostbite in the feet of turkeys, while at the same time, it prevents heat loss in the core of the turkey's body. Have you ever had a brain freeze from drinking too much of a slushy? Once a brain freeze hits you, what do you do? You stop drinking the thirty-two-degree slushy drink and wait for your body to warm up. That's what would happen to turkeys if they did not have the countercurrent heat exchange vein and artery design in their legs and feet. They would simply freeze to death in a matter of hours because their core body temperature would drop too much. Pretty cool, or fortunately not too cool. So I find myself really getting impressed by wild turkeys and wondered what kind of dinosaur they evolved from. Some experts in the field of evolution are suggesting that wild turkeys evolved from tyrannosaurus rex. I have trouble swallowing that idea. Whether or not this is true, it is not the key issue. What is paramount in the evolution of birds is that the consensus in evolutionary circles today is that birds truly evolved from dinosaurs. Do I concur with this hypothesis? No!

What does the theory of evolution require from dinosaurs that would allow them to evolve into birds? Everything! Birds have four chambered hearts, and dinosaurs have three chambered hearts. Birds are endothermic, warm-blooded, while dinosaurs are ectothermic, cold-blooded. Birds have feathers, and dinosaurs have scales. The lungs of birds are uniquely designed for a high metabolic rate that birds need to allow them to fly, while the lungs of reptiles and dinosaurs are very different. The bone structure of birds is very light, and some of the bones are hollow with cross supporting structures that add strength. Birds have wings and so on.

To summarize the evolution of a dinosaur to a bird, we need dinosaurs to grow wings, change their scales into feathers, add another chamber to hearts, and redesign their lungs and their entire metabolic system from cold-blooded to warm-blooded. We are asking a lot. Logically the process does not make any evolutionary sense. Every change in a living organism must have some type of beneficial trait that would allow that organism a better chance of survival, and these changes have to happen simultaneously to a male and female from random acts of genetic mutations in the DNA of both the male and female. Both the male (A) and female (B) would have to also find each other on this planet. Would the process start out with A and B sprouting baby wings with scales, then bigger wings with feathers, then a whole new lung design coupled with a new and improved four-chambered heart. All the individual steps that would involve changes in the DNA of a dinosaur would not become beneficial until you have a fully functional bird. If you really think about what the theory of evolution is asking you to believe, it is truly unbelievable, not logically and physiologically impossible. This type of scenario is repeated in every step of the way in the entire theory of evolution. It is one thing to say birds evolved from dinosaurs and quite another to prove it with fossil evidence and some degree of reasonable logic.

In college thirty-five years ago, I majored in geology, and I am very familiar with the science of evolution and the evidence and reasons behind that theory coupled with the idea that the earth is 4.6 billion years old. As a geology student, I believed in the theory of evolution, but at the same time, I believed that God was involved in the process. I certainly believed that the earth was billions of years old, and that the whole process of evolution was true, except that God guided that process. In my paleontology class, we were required to write a term paper on something related to evolution. I chose to write on creationism versus evolution. What I learned was that the entire theory of evolution is scientifically impossible, and the so-called missing links that are supposed to prove that evolution has occurred are being proven false time after time. There may be some missing fossil links out there, but I believe there are very few and possibly none.

Learning about the theory of evolution in school or college is a very one-sided story. Our nation has taken the idea of separation of church and state very seriously and disallowed virtually any discussion on creationism. Did God create the universe and all life on planet earth? Not if you go to school in America. My point. We are living in a time where getting to the truth requires some digging. Today the big story is fake news. Well, what is fake news, and who is the judge of what is true and what is not? That is the goal of this book. I want to convey the truth about the theory of evolution in a short but not overly technological book that will help serve as a stepping stone to the truth. Are we all random acts of chance that evolved for billions of years from a single cell bacterium in the primordial sea of earth's ancient world? Or did God create everything?

Chapter 1

IN THE BEGINNING

In the beginning, 13.8 billion years ago, in the vast empty expanse of space, there were no stars, no planets, no light, no gravity, nothing except an infinite amount of empty black nothingness called space. The time of 13.8 billion years is widely agreed upon by most astronomers. Then there was the big bang that started everything. This is already getting boring. I am going to my favorite place for a while.

Cape Cod! This is my favorite place. There is no other place in the world that means more to me than the memories I have of my family vacations in Cape Cod. To me and my family, the place is heaven on earth. The sun, the sand, the seals, the salty smell to the air all mixed with the relaxing sounds of wave after wave crashing on the shore. The sounds of seagulls crying far away adds to the symphony of sound if they aren't too close. Close-up, a seagull begging for food gets a little obnoxious. A rock tossed in the right direction at an obnoxious seagull that is too close retunes the proper ambiance for my Cape Cod soothing sounds on the beach.

My wife and I love to walk for miles upon miles on the beaches of Cape Cod. Fortunately, a very large portion of the cape has been forever preserved as a national seashore park, and no further development is allowed, except for a few grandfathered homes. So as you walk along the beach, all you see is sand and sand dunes and the ocean as far as the eye can see. It is always breathtakingly beautiful. Add to that the natural symphony of sound and pleasant smells, it truly becomes a place of peace and rest. Laying down on large beach towel, your hand naturally gravitates toward the sand, and you scoop up a handful and let it sift through your fingers. Each handful of sand contains over a thousand grains of sand. I wondered, *How many grains of sand are there in the world?* There are people out there who take the time to calculate such trivial ideas. The best answer that I found is that there is ten to the twenty-second power (that is a one with twenty-two zeros after it) grains of sand on the planet we call earth. The number of grains of sand that comprise all of Cape Cod are probably only one five thousandth of all the grains of sand on the planet.

That brings us to the question of how many stars are there in the universe. "The universe is all of time and space and its contents. It includes planets, moons, minor planets, stars, galaxies, the contents of intergalactic space, and all matter and energy."[2] The most accurate consensus among astronomers is ten to the twenty-fourth power, a one with twenty-four zeros after it. Comparing the total number of grains of sand on planet earth to the total number of stars in the universe turns out to be a relatively close comparison. The size of the observable universe is incomprehensible to our mind. It is estimated by astronomers that the average number of stars that a person can see with the unaided eye is between five thousand to twenty-five thousand stars. In the early nineties, I served six years in the National Guard, and I spent one summer training at Fort Irwin in Southern California. In the dry desert where there are no lights from civilization to hinder the view of the night sky, the number of visible stars is amplified tenfold. One night, we had a training mission, and we were up all night out in the desert, scanning the perimeter of our position using night-vision goggles. These devices amplify any faint source of light five thousand times. As I was using a pair of NVGs, as they are called, I laid on my back and scanned the heavens. What we can really see with our own eyes is truly nothing to what is really in our universe. I could easily see a hundred times more stars with the NVGs than with my own eyes. The heavens are so fantastically glorious when observed with night-vision goggles. Our planet resides with in the Milky Way galaxy, which is only one of an estimated ten billion galaxies in the universe. "A galaxy is a gravitational bound system of stars, stellar remnants, interstellar gas, dust, and dark matter. Galaxies range in size from dwarfs with just a few billion stars to giants with one hundred trillion stars, each orbiting its galaxy's center of mass."[3]

How big is the average star? Astronomers consider our sun to be an average-sized star. From planet earth, it looks about as big as a basketball, but how big is it compared to the size of earth? According to Universe Today, "1.3 million earths could fit into our sun."[4] Other estimates are just one million earths could fit into the sun. Either way, our sun, an average star, is at least one million times the size of earth. How large is the universe? The consensus among astronomers is that it is ninety-two billion light-years in diameter. Light travels at 186,000 miles per second, which means light can travel around the planet earth 7.5 times in one second. How big is a billion? If you count one number every second for twenty-four seven, it would take you over thirty-one years to count to one billion. This is our universe. It is utterly incomprehensible to our mind. This brings us back to the big bang theory.

> The big bang is the preeminent theory which attempts to explain the origin of the universe through purely natural processes. The theory points that all matter in the known universe started as a point of infinite density and temperature known as a singularity. It is believed that approximately 13.7 billion years ago, this singularity experienced a rapid inflation of matter, energy, space and time that eventually evolved and self-organized into stars, galaxies, and planets. The big bang was not an explosion in the conventional sense of the term, but an expansion of space and time. However, like an explosion, it was highly energetic and chaotic.[5]

This is not my theory; however, it is the best mankind has been able to come up with so far. The singularity is the heart of the big bang. Understanding what scientists are postulating when they talk about a singularity is important. One of the better definitions of singularity that I have found states, "In the center of a black hole is a gravitational singularity, a one dimensional point which contains a huge mass in an infinitely small space, where density and gravity become infinite and space-time curves infinitely, and where the laws of physics as we know them cease to operate."[6] To the best of my knowledge, the singularity in the big bang was as small as a pinhead or perhaps as large as a golf ball. Regardless, in the beginning, 13.7 billion years ago, all matter of the known universe (ten to the twenty-fourth power of stars) were compressed into that singularity. That's very impressive. I would like to see somebody stuff a five-gallon bucket full of sand into a golf ball.

I was getting my haircut one day at the local barbershop, and I picked up a science tech magazine to read as I was waiting to get my haircut. The article was discussing the big bang theory and the initial singularity that was comprised of matter that had infinite density and infinite temperature, combined with such an infinite gravitational force that even light could not escape. So if the gravitation was that great, what made the singularity blow up? It's a good question. The science tech magazine I was reading posed that same question. Their answer was that perhaps there was some antigravity matter present when the big bang blew up. I would like to know what antigravity matter is. Can I buy some for a smoother car ride? Maybe if I put a little bag of antigravity matter on the legs of my wild turkeys, they could fly better. What does infinitely dense and infinitely hot mean? How dense is infinitely? These are indeed perplexing questions. Still, as I read about the big bang theory, the question of what made the singularity blow up/expand is never adequately answered.

Another major question that I do not understand concerning the big bang theory is where did the singularity come from? Where did the matter come from that makes up every single star and planet in the known universe? How do you stuff all that matter into a golf ball? What is this big bang theory really stating? To me, it is stating that all the star and planets in the universe, each star a million times the size of planet earth, and each grain of sand on planet earth representing a star were all stuffed into a little golf ball-sized singularity of infinite density, temperature, and gravity. Then against all odds, it blew up. All I can say is wow, wow, big, big wow!

The big bang theory is known as cosmological evolution. It covers the origin of the universe all the way to the formation of our solar system and planet earth.

Again, the epic question about the origin of the universe is where did all that matter come from to make ten to the twenty-fourth power number of stars? As I have researched this, astronomers and evolutionist admit that they just don't know. They say science does not have all the answers yet.

As a geologist, I studied science and took many different science courses. The term *science* means "the systematic observation of natural events and condition in order to discover facts about them and to formulate laws and principles based on these facts."[7] There are numerous definitions of science. The word itself means knowledge. The heart of science is about employing the scientific method. When utilizing the scientific method, questions are asked about some aspect of our world, hypothesis are developed, predictions are made and tested, and conclusions are drawn. From the knowledge gained using the scientific method theories, laws are deduced. In the realm of science, theories are the best explanations of a tested hypothesis supported by a large body of testable scientific observations and data. Theories are the best answer to a hypothesis such as, how did the universe begin? Theories are subject to revision and modification as new data and scientific evidence arises. Scientific laws are a universally accepted as fact, not subject to change. "The universe obeys certain rules/laws to which all things must adhere. These laws are precise, and many of them are mathematical in nature."[8]

Having a basic understanding of scientific theories and scientific laws, it is important to note that the first law of thermodynamics contradicts the big bang theory. The first law of thermodynamics states that energy cannot be created or destroyed. According to Einstein's famous equation, $E=mc^2$, it is proven that energy and mass are interchangeable. So another way to state the first law of thermodynamics is to state that matter cannot be created nor destroyed. The laws of thermodynamic define how the universe works. These laws are the bedrock of all disciplines of science. If a scientific theory does not obey the laws of thermodynamics, it is breaking one of the most important of all known scientific laws.

Therefore, how is it that all the matter of the known universe can be created from nothing when the first law of thermodynamics states that this is impossible? Within the various definitions of the big bang theory, a clause that states that in the beginning of the creation of the big bang singularity, the known laws of physics did not apply. Convenient!

There is another theory about the origin of the universe. It starts out with "in the beginning, God created the heavens and the earth."[9]

Chapter 2

AN OLD EARTH?

Some people like the feeling of adrenaline and fear that you get when you do something crazy like cliff jumping off a thirty-five-foot cliff. One day, ten years ago, my fifteen-year-old daughter Vicky, my son Jeff, and I were jumping off a thirty-five-foot cliff into a virtually bottomless pool at a local swimming place called the Flume. A thirty-five-foot jump is scary. Of course, there's the fifty-foot jump about a quarter mile up the river from where we were currently at. Now fifty feet is more than I would ever jump. The problem with this fifty-foot jump is the landing. The gorge is V-shaped, and the river is only fifteen to twenty feet deep at this point if you are able to jump dead center into the river. The key is not ending up dead. I decided in my heart that I was absolutely not going to jump. We were all looking over the edge of this fifty-foot vertical cliff, which was enough to give me all the adrenaline I needed, when—*poof*—Vicky was gone. She jumped! My mind was considering the ramifications of this. My fifteen-year-old daughter just made a fifty-foot cliff jump, and Dad was too scared to jump. How many times for the rest of my life would this story be repeated?

This was not a time to reason too long, so I jumped. How was it? Just great! The sound of the air rushing past your ears as your body accelerates to forty-five miles per hour as it slams into water is not conveyable in words. You have to experience it. I intuitively knew that I was going too fast as I approached the water, so I spread out both my arms straight out from my body to slow down my plunge depth. Obviously, you should not do this if you want arms later in life, but when you are traveling at forty-five miles per hour into fifteen to twenty feet of water, you must decide if you would rather keep your legs or your arms. It sounded like a rifleshot as my arms hit the water, and I instantly plunged to the bottom of the V in the river. I allowed my legs to bend, which absorbed the rest of my downward momentum. My arms hurt for a week, but looking back at the decision of sticking out my arms, that may have saved my life and legs. I have never ever repeated this stunt. There was just too much fear and adrenaline coupled with a lot of stupidity for me to ever want to try this again.

You don't have to jump off a cliff to get that same kind of adrenaline rush and fear. Public speaking will also suffice. I took a public speaking class back in my college days, and I still remember one class. It was a group class, which had four public speaking classes that combined once a week in a huge classroom. The day's discussion was impromptu speaking. The professor asked for volunteers to try. Two crazy people volunteered and did PDG (pretty darn good). The professor asked for more volunteers, and the pool of crazies ran out. That meant a lottery-type draw from about 125 students. I was safe. The professor picked row seven and seat number six. Bingo. I was picked. I never win the lottery, but today was my lucky day. I still remember walking down to the front of the class. My heart felt like it was going to explode. I survived but did not ever want to repeat this experience. The fifty-foot cliff jump I have never again repeated, but in college, public speaking is an unavoidable facet of college life.

Do you remember the paleontology class that I took where I said I wrote a term paper on evolution versus creationism? Amazingly, I received an A from that professor who was a very firm atheist. The downside of this term paper was that I had to summarize and speak about my topic for fifteen minutes in front of my paleontology class. To the best of my knowledge, every student in this class was an ardent believer in the theory of evolution. The fear and adrenaline only lasted for about five minutes, and eventually I came to the end of my presentation and asked, "Are there were any questions?" Slowly, I looked all around the class, and no one raised their hand. Both the professor and I were shocked. The point of these reflections is that, here I was in a paleontology class with a group of students who were all studying the theory of evolution and after they were presented with evidence that creationism was a more viable explanation than the traditional Darwinian theory of evolution, the class was speechless. Why? Because schools and colleges just don't present any evidence that refutes the theory of evolution. At the time of my term paper presentation, I believed in what is called a "time gap theistic" take on creationism. I believed that God created all things, and that the biological evolution of one species into a completely different species simply did not occur. I did believe that the earth was 4.6 billion years old and that millions of years pasted between the creation of different species by God.

Since that time, nearly thirty-five years ago, I have spent a significant amount of time continuing my research on the creation/evolution debate. What I have learned is that there are many scientific truths and facts about the theory of evolution that school and college students will never hear. Please try to keep a somewhat open mind as I present these facts. Fact one, I now believe that the age of the earth is six thousand years old. Twenty years ago, I had a pastor who had just listened to a guy name Dr. Carl Baugh who taught that the earth was only six thousand years old, and he said he had the scientific evidence to prove it. I basically just laughed and said in my heart that this man was just ignorant. Today, I can say it was I who was ignorant, not him.

In geology, I was taught over and over again that the earth was millions and billions of years old. The evidence that I was presented in college geology classes seemed very logical. From a very young age, most of us are taught that the earth is millions of years old. Every single scientific program on animals or on planet earth, always stresses that this animal evolved from that animal millions of years ago. How long ago did dinosaurs live? Millions of years! It's a reflex preprogrammed answer. How long does it take for buried organic matter to turn into oil and natural gas? Millions of years.

Have you ever noticed all the candy cane-shaped pipes at our local landfills? They are there to allow methane and other gases to escape from the ground, which prevents the landfill from blowing up. In many places in our country, the methane that is produced from landfills is captured with elaborate pipe systems, collected, and then burned in generators, producing electricity at a local level. This methane gas produced at landfills is similar in composition to the natural gas found deep underground. Natural gas does have a higher methane concentration, but the gas from the landfill has enough methane concentration to burn it in generators, creating usable electricity. My point, right before our very eyes, we see natural methane gas continually being produced at the local landfills in a matter of years, not millions of years. Does this prove that the earth is young? No, but this is just one of many discrepancies of an old earth theory that I have come to understand over the years since my college education in geology.

Cenozoic, Mesozoic, Paleozoic, Precambrian are forever imprinted on my mind. It's kind of like the tongue twister, "Peter Piper picked a peck of pickled peppers. A peck of pickled peppers, Peter Piper picked." I memorized that tongue twister on a Cape Cod vacation forty years ago. The Cenozoic, Mesozoic, etc. names refer to geological eras that represent millions of years of time.

ERA		PERIOD	LIFE FORMS
CENOZOIC	2,6	QUATENARY	
	2,3	NEOGENE	
	66	PALEOGENE	
MESOZOIC	145	CRETACEOUS	
	200	JURASSIC	
	252	TRIASSIC	
PALEOZOIC	300	PERMIAN	
	360	CARBONIFEROUS	
	420	DEVONIAN	
	443	SILURIAN	
	485	ORDOVICIAN	
	544	CAMBRIAN	
	4,6 billion years ago	PRECAMBRIAN	

MILLIONS OF YEARS AGO

These geologic eras are further broken down into smaller sections of geologic time called periods. Looking at the above geologic time chart, it looks impressive and very exact. But is this geological timetable truly accurate? In the next page or so, I will discuss the basic foundational principles of geology that geologist use to determine the age of rock layers and the age of the earth.

The field of geology really began with James Hutton (1726–1797), who is known as the "founder of modern geology." At the time of his life, current reasoning about the age of the earth was based on genealogies of the time lines in the Bible giving the earth an approximate age of six thousand years old. As Hutton carefully observed the world around him, he wondered about the age of the earth. He looked at sedimentary rocks and how slowly they formed and reasoned that if the weathering and current rates of sedimentation that he was observing in his day were the same as in the past, sedimentary rocks and the geologic features that we see around us would take millions of years to form. "Sedimentary rocks are made up of pieces of preexisting rocks. Pieces of rock are loosened by weathering, then transported to some basin or depression where sediment is trapped. If the sediment is buried deeply, it becomes compacted and cemented, forming sedimentary rock."[10] He reasoned that by observing the formation of sedimentary sandstones, if one-fourth of an inch of sandstone takes one hundred years to be deposited, this knowledge could be extrapolated to estimate how long it would take to form a sandstone layer two feet thick. Therefore, if one-fourth inch of sandstone is deposited every one hundred years, a two-foot layer would take 9,600 years to be deposited. This theory became known as uniformitarianism, "The present is the key to the past." During the mid-1800s, Sir Charles Lyell, "the father of modern geology," expanded on the principle of uniformitarianism, promoting that theory in his series of books, *The Principles of Geology*. The idea of uniformitarianism is also widely known as gradualism. The geological changes that we observe today have all happened gradually over millions of years. The principle of uniformitarianism and gradualism are both stating the same idea. This new paradigm of geology changed the thinking of that time that the earth was six thousand years old. Suddenly the earth became billions of years old. This new "old earth" idea greatly influenced Charles Darwin, "the father of evolution," when he wrote the *Origin of Species* in 1859. Coupled with the principle of uniformitarianism is the law of superposition, which states, "In any undisturbed sequence of rocks deposited in layers, the youngest layer is on top and the oldest on bottom, each layer being younger than the one beneath it and older than the one above it."[11] Taking these three principles of geology and adding index fossils and radiometric dating, geologist are able to date individual layers of rock and ultimately the age of the earth.

Hutton, Lyell, and Darwin's theories at the time of their conception caused quite a stir in the Christian world view of the 1800s. At that time, the predominate Christian worldview was that earth was six thousand years old and that God created all life, man included, in six twenty-four-hour days. Man became so evil that God destroyed all life on earth with a global flood, saving Noah and his family. Noah built an ark, taking two of every kind of animal on board, and the world was later repopulated from those animals and Noah's family. I admit that Noah, the ark, and the global flood can sound like quite the tall tale. Of course, believing that a big bang created the universe and a one single-cell animal created all life on planet earth is also quite a tall tale. Which is true? Understanding the basic principles of geology will help clarify the question of how accurate is the geological timetable and the age of the earth. The true age of the earth is paramount to the creation/evolution debate.

Index fossils are "a widely-distributed fossil, of narrow range in time, regarded as characteristic of a given geological formation, used especially in determining the age of related formations."[12] The most common index fossils used today are *ammonites*, which lived 65–240 million years ago and are believed to have gone extinct when the dinosaurs also became extinct.

Trilobites are also a very common index fossil of arthropods that lived 521 million years ago until the end of the Permian period to 240 million years ago.

So when you find a trilobite in a layer of rock, based on geologic time reasoning, that rock must be older than 240 million years old and younger than 521 million years old. There are many other index fossils that look to me like a bunch of seashells, but I suppose they are not. When geologist use index fossils to determine the age of rock layers, they are assuming that the theory of organic evolution is true. That all life evolved from a single cell and slowly evolved into more and more complex life-forms. Therefore, when looking at the oldest sedimentary rocks on this planet, life-forms should at first appear very simple in complexity and, gradually over the course of millions of more years of evolution, become much more complex. During most of earth's four-billion-year history, evolution did not produce much more than bacteria, plankton, multicellular algae, and sponges up until about the six hundred-million-year mark. Shortly, thereafter, near the 540-million-year point of earth's history, the Cambrian explosion occurred. "The 'Cambrian Explosion' refers to the sudden appearance in the fossil record of complex animals with mineralized skeletal remains. It may represent the most important evolutionary event in the history of life on earth."[13] This event has been an enigma to geologist and evolutionist. Before the Cambrian Period, virtually all life-forms that existed were extremely simple, one-celled organisms.

Fossils are either the evidence of an old earth or a young earth, depending on a geologist or creationist interpretation. Take the Cambrian explosion, geologist and evolutionists really do not have a logical scientific reason as to why it occurred. Creationist say, "Perhaps God?"

What is a fossil? Fossils are "the evidence in rock of the presence of a plant or animal from an earlier geologic period. Fossils are formed when minerals in groundwater replace materials in bones and tissue, creating a replica in stone of the original organism or their tracks."[14] The critical question is, how long does it take a fossil to form? Creationist believe "a short time," while they state that geologists believe fossils take millions of years to form. That is incorrect. Geologist are completely aware that most fossils form quickly from local flooding caused from storms or hurricanes. Still the question is how long does it take to create a fossil, and what amount of sediment is necessary to create a fossil? There are very few cases of known fossilization occurring today. I live near a local river that periodically floods and deposits several inches to a few feet of silt and mud each year, but I have never discovered a fossil at this river. Never! I believe the key to fossilization is that when an organism is buried, it requires a lot of sediment two to ten feet or more in an ocean-type environment that will provide the minerals in the water needed to replace the materials in the organism to become a fossil. Rapid burial by significant sedimentation with the continued presence of water is the paramount.

Radiometric dating is "is any method of determining the age of earth materials or objects of organic origin based on measurement of either short-lived radioactive elements or the amount of a long-lived radioactive element plus its decay rate."[15] That sounds complex. What it means is that there are several radioactive elements that have a known rate of decay where the parent element will decay into the daughter element with a calculated half-life.

Some of the most common radioactive elements used today are as follows:

A. Potassium-40 to Argon-40 half-life 1.25 billion years
B. Uranium-238 to Lead-206 half-life 4.47 billion years
C. Carbon-14 to Nitrogen-14 half-life 5,700 years

Radiometric dating only works well on igneous rocks. Igneous rocks are the rocks that formed directly from liquid magma or lava. Once an igneous rock is formed, it is believed that the high heat has burned off all impurities out of the rock, leaving the parent radioactive elements to decay to their daughter elements. The amount of daughter element left in a rock is used in ratios against the parent element to figure out the age of the rock. The whole process of radiometric dating sounds like snake oil.

The question we need to ask is, does radiometric dating really work? Geologists say yes, and creationists say no. I thought a good test would be date volcanic rocks of known age. This has been done numerous times. The results produced multiple examples of where lava flows of known ages, less than two hundred years old, yielded radiometric dates of a few hundred thousand years old to two to three million years old. This has happened many times. These age discrepancies are common on dating rocks of known age of less than a few hundred years old. Geologists claim that these dating methods do not work on rocks that are young. They only work on old rocks. It seems reasonable to me that testing rocks of known age to verify that radiometric dating really works is logical and scientific. So if radiometric dating does not work on rocks of known age, how can we trust these dating methods to work on rocks of unknown ages? This gives me grave reservations in trusting the reliability of radiometric dating. This subject becomes very technical in a hurry, and there are many articles available that discuss radiometric dating on the web. I prefer to concentrate on evidence that is more logical, scientific, and verifiable like the sedimentary deposition rates in the Grand Canyon.

"The geology of the Grand Canyon area includes one of the most complete and studied sequences of rock on Earth. The nearly forty major sedimentary rock layers exposed in the Grand Canyon and in the Grand Canyon National Park area range in age from about two hundred million to nearly two billion years old."[16] The Grand Canyon is one of the seven great natural wonders of the world. It is 277 miles long and, at some points, is eighteen miles wide and up to one mile deep. The sedimentary layers vary from three thousand to six thousand feet thick. That means the exposed sedimentary layers in the Grand Canyon represent approximately 1.8 billion years of earth's most recent history. Looking at the plethora of iconic photographs of this canyon, the layers of individual sedimentary rock layers are very consistently flat and horizontal. The bedding planes, the line separating two distinct rock layers from each other, are also very straight and horizontal. This is significant evidence of what happened during the formation of these rock layers. Namely, nothing. If the sedimentary rocks of the Grand Canyon took 1.8 billion years to form, each time a different type of sedimentary rock layer appears, major environmental changes must have taken place.

Horizontal bedding of the Grand Canyon

Geologists know what conditions are necessary to form different types of sedimentary rocks. In the Grand Canyon, some of the layers are shales, sandstones, and limestones. Each different type of sedimentary rock layer requires a different kind of water environment to produce that type of rock. A shale requires a water environment that is full of mud and silt. Limestone is a sedimentary rock composed of 50 percent calcium carbonate, and it usually forms in clear shallow warm ocean waters. Sandstones are again a sedimentary rock comprised of silica sand cemented together with some type of cement such as silica or calcium. These are three very different types of rocks, each needing a very different source of material. My point is that to change the ocean environment from making shale with mud and silt, which would be a dark dirty water environment to clear water making limestone, is profound. There should be a time gap of tens of thousands to millions of years between different types of layers to allow the environment to change to produce a completely different type of sedimentary rock layer. This would produce significant erosion between the bedding planes leaving anything but the flat, horizontal bedding planes seen throughout the Grand Canyon. One other problem in the Grand Canyon is the Coconino Sandstone. Geologists believe that the Coconino sandstone was derived from a dry sand dune desert-type of environment. This has been debunked by Dr. John Whitmore[17] and other scientists. If you look at a picture of the white layer of Coconino Sandstone, it reveals a very consistent, evenly deposited horizontal layer of white sandstone. I do not know of any known dry desert with sand dunes in the world that could ever be deposited in such an even horizontal fashion, except if that sandstone was deposited by water.

Is the age of the Grand Canyon correct? Geologists claim that the ages of the sedimentary rock strata that are visible range from two hundred million years old to two billion years old with an average thickness of these layers being one mile. Therefore, we know that approximately one mile of sediments in the Grand Canyon required 1.8 billion years to be deposited. With that information, we can calculate the rate at which sediments in the Grand Canyon were deposited and compare those rates to known rates of sedimentation occurring in the oceans today. The principle of uniformitarianism declares the present is the key to the past; therefore, the rates at which sediments are deposited in the oceans should correlate to the same rates that sediments were deposited in the Grand Canyon. If the sedimentation rates do not match, then an age correction needs to be made to the Grand Canyon.

Within the ocean, there are vastly different sedimentation rates depending upon what part of the ocean you are in.

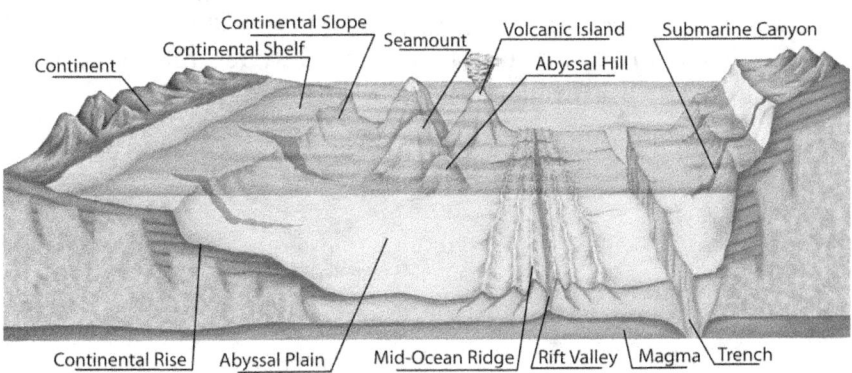

When geologists discuss what type of environment the rock layers of the Grand Canyon were formed in, the consensus among geologist is some type of shallow sea. Shallow seas are found on the continental shelfs. "The continental shelf is an underwater landmass which extends from a continent, resulting in an area of relatively shallow water known as a shelf sea."[18] Researching current sedimentation rates found in the oceans today, I found that "shelf sediments accumulate at an *average rate of 30cm/1,000 years*, with a range of 15–40centimeters/1,000 years. Though slow by human standards, this rate is much faster than that for deep-sea pelagic sediments."[19] According to Water Encyclopedia science and issues, "Sediment can accumulate as slowly as (0.4 inches) 1,000 years (in the middle of the ocean where only windblown material is deposited) to as much as 3.25 feet per year along coastal margins. More typical deep-sea rates are on the order of *several centimeters per 1,000 years*."[20] Utilizing this current scientific data on average sedimentation rates of in the oceans, I will use the average rates for continental shelves, as well as the average sedimentation rates for the deep ocean.

1. First, we need to calculate the sedimentation rate of the Grand Canyon sediments. One mile of Grand Canyon sediment is equals to 160,934 centimeters/1,800,000,000 (1.8 billion years) is to .0000894077 centimeters per year.
2. For one thousand years that sedimentation rate in the Grand Canyon is .0008940777 × 1,000 = .0894/1,000.
3. For ten thousand years, that rate is .894 centimeters. That means that according to the proposed ages of the Grand Canyon sedimentary rocks that every ten thousand years, almost 1 centimeter of material was deposited.

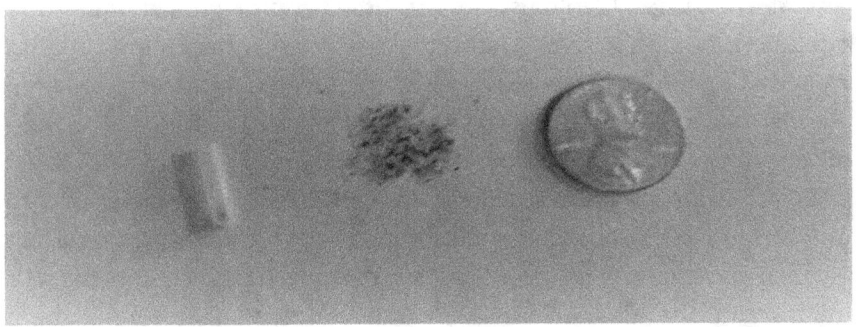

This is what a centimeter of fine-grained sand looks like. On the far left of this picture is a one-sixteenth-inch diameter plastic tube cut to one centimeter in length. I filled it with fine-grained sand and emptied it out next to the penny. It took about five hundred grains of fine-grained sand to fill it. That means on the shallow ocean bottom of where the Grand Canyon sediments were being deposited, once every twenty years, a fine-grained piece of sand was deposited on the entire surface of the bottom of that shallow ocean floor. This is an unreasonably slow sedimentation rate. Common sense should tell any geologist that this sedimentation rate is not realistic.

1. The average sedimentation rate in the deep oceans is 2 cm/1,000 years. Taking two and dividing it by one thousand years will give the annual sedimentation rate (2/1000 = .002cm per year).
2. Dividing the depth of the Grand Canyon layers again, 1 mile = 160934 cm/.002 = 80,467,000 years. That is eighty million years.
3. Using average sedimentation rate for continental shelves of 30 cm/1,000 years, we divide 30/1000 = 0.03cm/year.
4. Depth of the Grand Canyon sediments, 1-mile 160934cm/.03 = 5,364,466 years. Rounded off, that number is 5.4 million years.

Five point four million years old is what the age of the Grand Canyon should be based using the current sedimentation rates found on continental shelves. Geologists might disagree with the sedimentation rate that I used of 30 cm/1,000 years, which gives the Grand Canyon an age of 5.4 million years. Utilizing the sedimentation rate of deep oceans at 2 cm/1,000 years. That would increase the age to 80.5 million years old for a pile of sedimentary rocks that are supposed to be 1.8 billion years old. Taking 1.8 billion and converting that to millions is 1,800 million. The difference between the uniformitarian dates of the Grand Canyon sedimentary rocks using the average sedimentation rate of 30 cm/1,000 years yielding an age of 5.4 million minus 1,800 million is equals to 1,794.6 million. That means the age of the Grand Canyon is off by 1,794.6 million years. Using deep-ocean sedimentation rates, the age of the Grand Canyon is still off by 1,719.5 million years.

Some geologist may find fault with the thickness of the Grand Canyon layers that I used to calculate the correct age of the Grand Canyon, stating that there are missing sedimentary layers in the Grand Canyon, and the true thickness of the 1.8 billion years of Grand Canyon sedimentary layers should be much thicker than one mile. That is a legitimate complaint, so I will double the thickness of the Grand Canyon to two miles. New calculations: 2 miles is 321,868 cm/.03 equals 10,728,933, or 10.7 million years old. No matter how an age is calculated for the Grand Canyon's age based on current sedimentation rates found on the continental shelf or deep ocean, the error of the Grand Canyon's age is vastly incorrect by over 1,700 million years using the uniformitarian principle of geology. Sorry.

Of course, this does not consider the greater sedimentation rates required to burry all the fossils found in the sedimentary rocks of the Grand Canyon. *Comparing the current rates of sedimentation in our oceans to how long would it take to lay down one mile of sediments is concrete evidence that the entire age of the Grand Canyon is off by at least 1,700 million years. Not a small error.* If the age of the Grand Canyon sedimentary layers should be five to ten million years old, bearing in mind that there is virtually no other place on earth where the geology of rock layers has been more thoroughly studied, then what should the true age of the earth be? The answer, less than twenty-five million years old based on the current sedimentation rates found on continental shelves, using the uniformitarian principal, "the present is the key to the past," as our time ruler. These conclusions are not speculation. I think the age of the Grand Canyon is way off because it does not fit my world view. No, I employed simple math and simple logic and found that if geologists are going to use the uniformitarian principle as a time ruler, current rates of sedimentation on the continental shelf, shallow seas, should have matched the Grand Canyon rates of sedimentation. They did not. If the principle of uniformitarianism produces discrepancies in the age of the Grand Canyon sedimentary rocks that are 90 percent or more too great, where in the world does the uniformitarian principle work?

There are many other enigmas the uniformitarian principal of geology cannot explain. A polystrate fossil is a fossil of a single organism (such as a tree trunk) that extends through more than one geologic stratum.[21] All over the world, such fossilized trees can be found extending to five, ten, to forty feet through different layers of sedimentary rocks. If we use the sedimentation rate of the Grand Canyon of one centimeter of sediment deposited every ten thousand years, a twenty-foot-long piece of tree extending through multiple layers of sedimentary rock layers would equate to over six million years of sediment. Certainly, this is not reasonable or logical. What all polystrate fossils represent is proof that the surrounding sediments were deposited very rapidly. Geologists have an answer for these fossils. A local flood. However, when the sedimentary rocks that polystrate fossils are buried in extend for many, many, miles, the idea of a local flood is not adequate to explain that polystrate fossil.

The earth's magnetic field has been measured for the last 150 years. Scientists have discovered that the field is decreasing. "The average intensity of the Earth's magnetic field has decreased exponentially by about 7 percent since its first careful measurement in 1829." With this knowledge, creation scientist extrapolate the increased strength of the earth's magnetic field back in time and conclude that beyond ten thousand years in the past, the strength of the magnetic field would be too great for animal life. Geologists disagree with the magnetic field continuously decreasing at a constant rate for the past and claim that the earth's magnetic field fluctuates and poles experience magnetic reversals from time to time. Neither creationist nor evolutionists can absolutely prove whether the magnetic field is constantly or not constantly decreasing. Certainly, current observer's date proves that it has decreased since it was first measured. Apparently, "the present is the key to the past" does not apply with the earth's magnetic field.

Within the fields of geology and evolution, there are two words that I have never heard discussed in any of my college classes; they are *fossil graveyards*. They are taboo words to paleontologist. Stating those words to a paleontologist is like tossing a cross to a vampire. Picture the weather channel and one of their winter storm pictures of a fifty-car pileup on some highway. There is nothing but devastation, crushed trucks, and cars all shoved this way and that way and death. That is a fossil graveyard on a microscale. All over the world and in our country, there are massive deposits of animals, fish, dinosaurs, sharks, elephants, wooly mammoths, whales, and most other living creature that look like they were all dumped out of a C-130 airplane, flying at ten thousand feet. They are places of epic animal massacres that revealed a quick and violent death for herds of different species of animals, seeking to escape what looks like the end of the world. These fossil graveyards range in size from a few acres to forty square miles and larger. They can be found in North America, South America, Europe, Asia, Australia, Africa, and all over Antarctica, which represents all the world's continents. In America, they can be found in Wyoming, North Dakota, South Dakota, Florida, South Carolina, California, Nevada, Arizona, Texas, Idaho, Alaska, and many other states.

Here's a quote from a California fossil graveyard, "The famed Shark tooth Hill Bone Bed in California is loaded with shark teeth as big as a hand and each weighing a pound, from giant prehistoric killers called megalodon. Intermixed with copious bones from extinct seal, whales and fish, as well as turtle shell three times the size of today's leatherback, all these relics seem to tell of a fifteen-million-year-old disaster."[22] In South Carolina, there is another megalithic fossil graveyard named the Ashley Beds that was documented by F. S. Homes (paleontologist and curator of the college of Charleston's

Natural History Museum), and he described the fossil graveyard in a report to the academy of natural sciences, "Remains of the hog, the horse and other animals of recent date, together with human bones mingled with the bones of the mastodon, and extinct gigantic lizards."[23] This book was written in 1870 when the term *dinosaur* was not commonly used. The word *dinosaur* means terrible lizard and was first coined in 1841. The Ashley beds also contained whales, sharks, purposes, elephants, deer, sheep, fish, and dogs all intermittingly thrown together in a giant fossil graveyard that extends for miles. This fossil graveyard is another example of some type of horrendous environmental disaster that occurred in the past. In fact, all the fossil graveyards found in every corner of the world represent some type of cataclysmic disaster from the past.

I live at an elevation in Wilmington, New York, that Google Maps confirms is 1,093 feet above sea level. I have two daughters who are attending college at Sierra Nevada College located next to Lake Tahoe, and the elevation there is 6,350 feet above sea level. I have a son in Colorado Springs, Colorado, and the elevation in that city is 6,030 feet above sea level. My oldest daughter lives in Salt Lake City, Utah, and the elevation there is 4,226 feet above sea level. They all should be safe if climate change causes all the ice on the planet to melt, including the North Pole, Antarctica, Greenland, and all the world's glaciers. I wondered how much the oceans would rise if all the ice on the planet melted. Specialist in such questions states, "There are over five million cubic miles of ice on our planet. That includes the Artic, the Antarctic, and a swath of ice sheets and glaciers crusted around Earth's northern and southern boundaries. If it were to melt, global sea levels would rise at least 216 feet."[24] I have read other estimates as high as 260 feet. I believe 250 feet is a realistic upper limit to sea level rises if all the world's ice melted. This map of North America is what our continent would look like if the all the ice in the world melted.

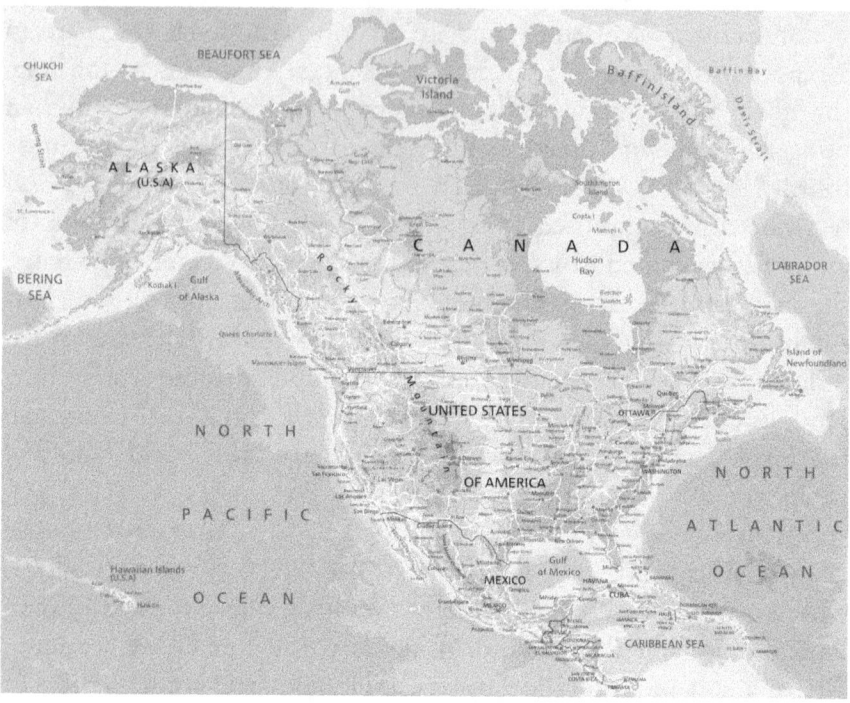

We would lose Florida for sure. Cape Cod, my families' favorite vacation spot, would disappear. That would be terrible. Looking at the map, what do you see? Most of the United States is not underwater. Geological surveys indicate that "sedimentary rocks have an average thickness of 1,800 meters on the continents."[25] "The sedimentary rock cover of the continents of the earth's crust is extensive (73 percent of the earth's current land surface)."[26] Where did the oceans of our planet get enough water to rise and fall covering 73 percent of the land surfaces of all the world's continents with thousands of feet of sedimentary rock? Everything in geological teaching regarding the history of sedimentary rocks, especially the Grand Canyon, always teaches how millions of years ago, there was some type of shallow sea that deposited that sedimentary layer of rock. A shallow sea means water covered most of the continents at some time in the past. If we melt all the ice in the world, we will only raise the level of the oceans by 250 feet. I see this as a serious problem. I understand in geology that there are events where mountains are uplifted by continental collisions and, at various times in the past, portions of continents were lower than they are today. The elevation of the Colorado River in the bottom of the Grand Canyon is 2,200 feet above sea level, and the top of the rim averages nearly seven thousand feet. Yes, the sedimentary rocks of the Grand Canyon were uplifted, but was the entire area ever at an elevation of less than 250 above sea level? There is only so much water on our planet's surface. The atmosphere can only hold so much water. For oceans to rise and fall, the only viable source is melting the Antarctic and the rest of the world's ice formations. Scientist can certainly calculate within plus or minus twenty-five feet, the maximum amount oceans can rise if all the world's ice melted. So how is it scientifically possible that shallow seas covered most of the world's continents in the past based on the current amount of water available today?

That raises the question of where did all the water come from that caused the Noah flood? Certainly, we know as a fact that sedimentary rocks cover most of the continents to an average depth of 1,800 meters, which indicates that at one time, water covered most of the continents. So where did the extra water come from? Evolutionists have a document that they hold in very high esteem which tells them how life evolved from a simple cell to humans for over four billion years, *On the Origin of Species* by Charles Darwin. Geologists have created a document called the geological timetable based primarily on a single principle, "the present is the key to the past," coupled with the tenants of Darwin's theory of evolution, which geologists and evolutionists use to interpret the ancient history of earth. Creationists also have a document that tells them about the history of earth called the Bible. "A new Harris Poll finds that a strong majority (74 percent) of US adults say they believe in God."[27] In 2013, a poll by Rasmussen Reports found that a majority of all Americans (63 percent) believe that the Bible is literally true and the word of God" (CBS News.com, 2013).[28] I found in other polls that the number of Americans who believe that the Bible is the literal word of God is closer to 30 percent. The point is, most Americans believe in God, and millions believe that the Bible is literally true. I do. When I was a young boy, I very distinctly remember lying on my bed and trying to decide if God was real. As I lay in bed, one of the arguments that I pondered in my head was, *If God is not real, how is it that there are so many thousands of churches all over America with the millions of people that attend church?* I reasoned, *How could so many people go to church if there was no God?* That was a point in my life over forty-five years ago that I decided that there must be a God. Today, I reason that if God is God and he created the universe and all living creatures on earth and man and woman, then he should be smart enough to write one book to his creation, the Bible. The question now is, can we believe that the Bible is literally true cover to cover? So far, I have presented the best evidence mankind has been able to think of concerning the creation of the universe and the age of the earth. An old earth based on a uniformitarianism philosophy that does not match the current rates of sedimentation in our oceans to the historic rates of the Grand Canyon by 1,700 million years does not account for how the oceans in the past were able to cover most of the continents when an upper limit of a 250-foot maximum rise in the level of the world's oceans is all that is scientifically feasible, and it certainly does not account for the plethora of mass fossil graveyards. In college, I believed most of what scientist taught about an old earth and the theory of evolution, but because of the knowledge I have learned since then and thus far discussed, I find that mankind's best scientist do not have the answers.

    The global flood. How big was it? The Bible says it covered the tallest mountains by over fifteen cubits. A cubit is about twenty inches long, which would make the depth that the water covered the top of the tallest mountains, close to twenty-five feet. That's a lot of water. Where did it come from? The book of Genesis in the Bible tells some very interesting facts about the history of our planet. In the time before the flood of Noah, the Bible states in Genesis 2b (KJV), "For the Lord God had not caused it to rain on the earth, and there was no man to till the ground; but a mist went up from the earth and watered the whole face of the ground."[29] Most Bible scholars agree that before the flood of Noah, it did not rain on the earth, but instead plants were watered from a layer of underground water that caused mist to rise and water all the living vegetation. In Genesis 1:29–30 (KJV), the Bible states, "And God said 'See, I have given you every herb that yields seeds which is on the face of the earth, and every tree whose fruit yields seed; to you it shall be for food. Also, to every beast of the earth, to every bird of the air, and to everything that creeps on the earth, in which there is life, I have given every green herb for food;' and it was so."[30] Before the flood of Noah, man and all living animals were vegetarians, and it did not rain. Another interesting fact in Genesis is the genealogy ages listed in the Bible. Genesis 5:4–7 (KJV) states, "And the days of Adam after he had begotten Seth were 800 years: and he begot sons and daughters: and all the days that Adam lived were 930 years: and he died, and Seth lived 105 years, and begat Enos: and Seth lived after he begat Enos 807 years, and begat sons and daughters."[31] And so on. It is the Jewish tradition that Moses wrote the book of Genesis. Think about this, Moses lived many years after the flood, and in Exodus 33:11, the Bibles says, "And the Lord spoke unto Moses face to face, as a man speaketh unto his friend."[32] When Moses wrote the book of Genesis, he got the facts straight from the source, which means Genesis is accurate. If the Bible is just made-up imaginations from the hearts of men, who could have ever conceived such a story as Genesis? Why would anyone make up a story that for the first 1,556 years of earth's history, men lived up to almost a thousand years old, were vegetarians, and it never rained, and the author of the story spoke face-to-face with God? Either it is the greatest imaginary story ever told, or it is true. "Modern archeology has challenged the world of education to admit that the Bible is factual. Solid, documented evidence outside the Bible record confirms events and persons that were at one time considered to be suspect or plain false."[33] One of the most famous stories in the Bible is when Moses led the Israelites across the Red Sea when the water parted, and when the sea closed back, it drowned the Egyptian army with all their chariots. These chariots have been found. Michael Rood, a Hebrew Roots teacher, has produced a video titled, "The Red Sea Crossing." "What they found strewn across the bottom of the Red Sea has shaken the religious and scientific community," says Roots. "Cameras mounted on remote controlled submarines revealed coral encrusted chariot parts, horse and human remains strewn like a battlefield wreckage on the bottom of the Red Sea."[34] These chariot artifacts have been proven to be from the Egyptian's eighteenth dynasty. The cities of Sodom and Gomora have been found along with the surrounding cities of the plain. "Each of the cities of the plain contain evidence of brimstone which God rained down upon the cities to destroy them. We're talking millions of brimstone balls. The brimstone is composed of 96–98 percent sulfur with trace amounts of magnesium which create an extremely high temperature burn."[35] In nature, natural sulfur is only 40–45 percent pure.

The field of scientific archeology continues to prove that the stories in the Bible are not stories but historical fact. It is one thing to say I believe the Bible is true, but when archeology constantly proves that the history and stories in the Bible are true, it shows with great confidence that the Bible is much more than a book of faith; it is a book of historical facts.

THE FLOOD OF NOAH

> Then the Lord saw that the wickedness of man was great in the earth, and that every imagination of the thoughts of his heart was only evil continually. And it repented the Lord that he had made man on the earth, and it grieved him at his heart. And the Lord said, I will destroy man whom I have created from the face of the earth; both man and beast, and the creeping thing, and the fowls of the air; for it repenteth me that I have made them. But Noah found grace in the eyes of the Lord.[36]

The Lord then commanded Noah to build and ark of gopher wood that was approximately 450 feet by 75 feet by 45 feet and fill it with two of every kind of animal with food for them and his household.

> In the six hundredth year of Noah's life, in the second month, the seventeenth day of the

month, the same day *were all the fountains of the great deep broken up, and the windows of heaven were opened.* And rain was upon on the earth forty days and forty nights. And the flood was forty day upon the earth; and the waters increased, and bare up the ark, and it was lifted up above the earth. And the waters prevailed exceedingly upon the earth; and all the high hills, that were under the whole heaven, were covered. (Genesis 7:11–12, 7:17, 7:19 KJV)[37]

Approximately eleven months later, the flood was over, and Noah, his family, and the animals left the ark. This is what created all the sedimentary layers of rock that covers 75 percent of the continents to an average depth of 1,800 meters or (1.1 miles) and left the dead animals and plant life in great deposits that became fossil graveyards and coal found all over the world. The source of the water for the flood was a subterranean layer of water that was under the surface of the entire earth. The vegetation on the planet was watered from an underground water source coming up as mist. Creation scientists have also theorized that there was some type of water canopy that surrounded the earth, which would have increased the oxygen concentration of our atmosphere while at the same time, filtering out virtually all ultraviolet rays from the sun. This created a perfect earth that allowed man and animals to live for a close to one thousand years. After the flood of Noah, the age limit that man lived quickly decreased to less than one hundred years. In Genesis 1:6–7, this canopy is described, "Then God said, 'Let there be a firmament in the midst of the waters, and let it divide the waters from the waters." Thus, God made the firmament and divided the waters, which were under the firmament from the waters which were above the firmament, and it was so. And God called the firmament heaven. And the evening and the morning were the second day (NKJV).[38] The next two verses discuss how God created the dry land and called it earth. This information gives us two sources for the water of Noah's flood. There must have been massive amounts of water in a subterranean layer under the crust, and there was water in the canopy that surrounded the earth. When the fountains of the great deep opened, creation scientists believe that the fountains of the great deep burst forth out of the earth with such tremendous pressure that they ruptured the canopy of water above the earth, causing rain for forty days and forty nights. When the waters of the flood receded, they went back into the subterranean layer within the earth. "After decades of searching, scientists have discovered that a vast reservoir of water, enough to fill the Earth's oceans three times over, may be trapped hundreds of miles beneath the surface, potentially transforming out understanding of how the earth was formed."[39] There are quite a few articles on this subterranean water trapped deep within the earth published during the last few years and can easily be found on the web. Once again, we have scientific evidence providing validation for the source of the water in Noah's flood.

Chapter 3

SUB-THEORIES AND FORMATION OF EARTH

I love watching the way the sun shimmers off the ocean's surface on a warm summer day in Cape Cod, standing on the back deck of our ocean-view vacation rental. The clouds that float intermittently across an aquamarine sky cast shadows on an ocean that is one-minute blue, then green, sparkling, and then a mixture of all three. As the sun begins to set, the sky turns into a kaleidoscope of bright orange, flaming yellows, and ruby reds that is utterly mesmerizing. The yellow ball of the sun as it begins to sink below the horizon becomes smaller and smaller until only a sliver remains—that too quickly appears to be snuffed out like a candlewick being quickly squeezed between two fingers. That experience was heaven on earth during one of my family's vacation to Cape Cod in 2008. It was the only time we could afford to rent a vacation home on the Cape with an ocean view. There are a lot of people who have never even seen the ocean. They are missing out on one of life's greatest outdoor experiences. Looking off the deck of our vacation home at the ocean, the human eye sees nothing except water to the end of the horizon, where the curvature of the earth begins. When a ship enters the horizon, only the top of the ship is visible at first until it passes above the point of earth's slight curvature, then gradually more and more of the ship becomes visible. As I gazed at the ocean, I thought to myself, *There is a lot of water out there in all the world's oceans.*

> According to the US Geological Survey, there over 332,519,000 cubic miles of water on the planet. A cubic mile is the volume of a cube measuring one mile on each side. It's hard to imagine, but about 97 percent of the Earth's water can be found in our oceans. Of the tiny percentage that's not found in the ocean, about two percent is frozen up in glaciers and ice caps. Less than one percent of all the water on Earth is fresh. A tiny fraction of water exists as water vapor in our atmosphere.[40]

Evolutionists theorize that somewhere out there in the vast reaches of the world's oceans, the first cell arose 3.8 billion years ago. This facet of the theory of evolution is called chemical evolution. It is "the formation of complex organic molecules from simpler inorganic molecules through chemical reactions in the oceans during the early history of the earth."[41] Evolutionists believe that by random chemical processes given enough time, the first living cell arose by chance. The creation of the universe, stars, galaxies, and all the planets with their moons is a subdivision of the theory of evolution known as cosmological evolution. In years past, I have studied the complexities involved in the formation of the first simple cell that formed somewhere in the world's oceans, but I never gave much thought to where the earth's oceans came from, or better yet, where our planet came from, what caused it to form along with our moon? If one is to believe that life arose from a simple cell, it is important to understand the sub-theories involving the creation of our planet, the moon, and the 332 million cubic miles of water that fill earth's oceans. These sub-theories I will briefly expound upon, but the heart and soul of the theory of evolution is the creation of that first living cell from nonlife. This part of the chemical evolution needs much illumination. For if it is impossible for a simple cell to arise by chemical processes and chance, is it not the theory of evolution a moot point?

The most common theory that has been proposed for the birth of all the planets in our solar system is what is called the disk theory.

> Basically, scientists have ascertained that several billion years ago out Solar System was nothing but a cloud of dust particles swirling through empty space. This cloud of gas and dust was disturbed, perhaps by the explosion of a nearby star (a supernova), and the cloud of gas and dust started to collapse as gravity pulled everything together, forming a solar nebula—a huge spinning disk. As it spun, the disk separated into rings and the furious motion made the particles white-hot. The center of the disk accreted to become the Sun, and the particles in the outer rings turned into large fiery balls of gas and molten-liquid that cooled and condensed to take on solid form. About 4.5 billion years ago, they began to turn into the planets that we know today as Earth, Mars, Venus, Mercury, and the outer planets.[42]

Scientist tell us that the temperature of outer space is -455 degrees Fahrenheit, so I wondered how particles swirling through empty space could get so hot. I think of this theory as being comparable to a giant record player with different size marbles representing the sun and the planets spaced at appropriate distances. In college, thirty-five years ago, I had a record player, but today they are just a faint memory. This analogy may not compute for the younger generation. Once the sun formed, the dust swirled around it in a concentric disk like a record spinning on a turntable, and gradually the dust accreted into small rocks that grew larger and larger, eventually becoming our planets. For this theory to work, scientist know that in a solar system, "all planetary orbits should be roughly circular, and with minor exceptions, everything in a star system should orbit in the same plane and in the same direction."[43] However, today there are new state-of-the-art telescopes, one is the Kepler telescope. It is a space observatory that was launched into space in 2009. This telescope has discovered over one thousand different planets orbiting stars in our galaxy. But these planets' orbits do not logically fit into planet formation theories such as the disk accretion theory. According to National Geographic news, "About one in three exoplanets' orbits are misaligned. Some orbit in the opposite directions as their stars' rotations, and other are tilted out of the ecliptic, like weather satellites crossing over Earth's Poles rather than the Equators."[44] These new discoveries negate what was hailed as the best theory for planet formation. In the same article from National Geographic news, it states that "the more new planets we find, the less we seem to know about how planetary systems are born, according to a leading planet hunter."[45] The name of this article was "Three Theories of Planet Formation Busted, Expert Says." To the best of my knowledge, experts do not have any new planet formation theories that will account for planets whose orbits are tilted and opposite to their stars' rotations.

I like the formation of the moon theory.

> It was also during this eon, roughly 4.48 billion years ago (or 70–110 million years after the start of the solar system), that the earth's only satellite, the moon, was formed. The most common theory, known as the "giant impact hypothesis" proposes that the moon originated after a body the size of Mars (sometimes named Theia) struck the earth a glancing blow. The collision was enough to vaporize some of the earth's outer

layers and melt both bodies, and a portion of the mantle was ejected into orbit around the earth. The ejecta in orbit around the earth condensed, and under the influence of its own gravity, became a more spherical body: the moon.[46]

This is how the moon formed. I can't help but think of Frankenstein and his flat head when I ponder this theory. Astronomers have calculated the orbital speeds of planets around our sun and their speeds vary. The speed for Mars was estimated to be just under 54,000 miles per hour. In 2009, one of my business vans was involved in a head-on collision with a full-size truck. Both vehicles were traveling at close to fifty-five miles per hour. The devastation was surreal. I never in all my life have seen such wrecked vehicles from a car accident. Fortunately, no one was killed or seriously injured. In the moon creation theory, we have a small planet that struck earth a glancing blow at close to 54,000 miles per hour. As for a small planet, Mars is approximately half the size of earth and 15 percent of the mass of earth. I would think the result of such a collision would have shattered the earth and the other planet. Who knows? What I really do not understand is why the earth is still so round. If a planet hit our planet at 55,000 miles per hour, shouldn't there big a large missing junk of earth, kind of like Frankenstein's head?

Chapter 4

EARTH'S WATER

Our planet is known as the water planet. Seventy-one percent of earth is covered with water, with 97 percent of that water found in the oceans. Where did the 333,000,000 million cubic miles of water come from? Scientist have hypothesized three different possible theories that attempt to solve this question. Two of the most popular theories involve earth's water being deposited from either comets or asteroids hitting the earth. Asteroids contain 10–20 percent of water locked into their rocks, while comets contain about 80 percent water in the form of ice. Here is a quote about the asteroid theory from science news, "This bombardment of asteroids a few million years after the start of the solar system could have easily delivered enough ice—locked inside the rocks, safe from the sun's heat—to account for earth's oceans, computer simulations indicate."[47] For many years, the most popular theory of the two was that comets were the main suppliers of earth's oceans. If this theory was correct, the chemical signature of the water found in comets should be similar to the chemical signature of water found in earth's oceans. In 2004, the European Space agency launched a space probe called the Rosetta into space with a primary goal of landing Philae, a small lander on the comet known as 67P. Philae did land on the comet, but unfortunately it got wedged into a large crack in the dark side of a small cliff. The lander ran out of solar power in two days. Still, great quantities of data were achieved on this mission.

> New measurements from Rosetta, which is studying comet 67P/Churyumov-Gerasimenko, reveal that the chemical signature of water in the comet is nothing like what's found in earth's oceans. The discrepancy suggest that comets did not bring water to earth and the more likely source was asteroids, says planetary scientist Kathrin Altwegg, of the University of Bern in Switzerland.[48]

It is quite amazing that mankind actually landed a small spacecraft on a comet. Nonetheless, the data from comet 67P and other comets reveals that the chemical signature of water found on comets is too different from the chemical signature of the water in earth's ocean. This new information basically rules out the hypothesis that comets delivered most of earth's water.

That leaves us with asteroids and theory three that water formed with the earth. Asteroids do not contain nearly as much water as comets, typically only 10–20 percent of an asteroid contains water. That water that is locked into the asteroid's rocks. Nuts and Bolts. I need to digress on a relevant tangent for a moment. One of the greatest faults I find with the theory of evolution is contained in the expression, "Nuts and bolts." This expression means the basics of something or the practical facts about a particular thing rather than theories or ideas about it. The theory of evolution is full of sub-theories like where did earth's water come from or a simple cell, but rarely are the nuts and bolts of these sub-theories ever discussed in detail. I like to see the practical side of each of evolution's sub-theories. When I used simple math to determine if current continental sedimentation rates would match the proposed 1.8-billion-year-old age of the sedimentary rocks in the Grand Canyon, they did not match by 1,794.6 million years. It is an easy thing to say scientists believe earth's water came from asteroids, but it is the nuts and bolts of such a hypothesis that determine if such an idea has practical scientific merit.

Let's look at the nuts and bolts of the hypothesis that earth's water was supplied by asteroids. How much water would it take to fill earth's oceans? According to the US Geological survey, 333,000,000 cubic miles of water. That number was rounded off to simplify the following computations. That number represents one-third of a billion cubic miles of water. Asteroids and meteorites are classified by size. A meteorite is less than one meter in size whereas an asteroid is greater than one meter in size. Asteroids rang in size from one-half mile up to 590 miles in size. Currently, NASA has not calculated an average size for asteroids. It is believed by many scientists that an impact crater found near the town of Chicxulub in the Yucatan Peninsula, Mexico, was caused by an asteroid that was six miles wide. This asteroid is believed to have caused the extinction of dinosaurs sixty-five million years ago. The devastation caused by a six-mile-wide asteroid is unbelievable. "The explosion that created the crater, which is more than 110 miles wide, likely involved a hit from an object about six miles across. The crash would have released as much energy as one hundred trillion tons of TNT or beyond a billion times the power of the atom bombs that destroyed Hiroshima and Nagasaki" (Space.com).[49] The key to asteroids suppling earth's water is we don't want them larger than a mile, preferable one-half mile in size. Hypothetically, let's estimate that all the asteroids that supplied earth's water were exactly one cubic mile in size. That means we would need at least 333 million and one cubic mile asteroids. However, the available water content in asteroids is only 10–20 percent. For this discussion, I will use the 20 percent water content of asteroids as an example. This means that to get one cubic mile of water from asteroids that is one cubic mile in size, we will need five asteroids to get one cubic mile of water. Therefore, 5 times 333 million equals 1,665 million (1.65 billion) cubic miles of asteroids are needed to bombard earth to fill up earth's oceans. Eighty percent of the 1,665 million asteroids will be waste—asteroid debris. Where is that asteroid debris going to go? The volume of the earth's oceans will be completely filled with the first 333 million asteroids, then the debris from the remaining 1,332 million asteroids will cover the rest of the planet with the equivalent of four oceans full of asteroid rocks. Is our planet full of asteroid rocks? No! Of course, if the average size of asteroids hitting earth to fill our oceans is only one-half cubic mile, we would need 3330 million or (3.3 billion) asteroids to accomplish this. This is what I mean by nuts and bolts of a sub-theory. Scientist can easily say we believe the water in earth's oceans was delivered by asteroids, but if you do some simple calculations and think about the theory logically, it is not logical. The illogical part of the asteroid theory is what happens to the remaining 80 percent of asteroid rocks—debris? It must go somewhere. What about needing 1,665 million one cubic mile wide asteroids to strike earth. That's a lot of asteroids. The number is way beyond what is reasonable. If the surface of our planet was full of rocks that were composed primarily of the same materials as asteroids, I would agree that the asteroid water delivery theory has some merit. This is not the case.

There is one final theory that evolution scientists have proposed for earth's water source. That water was present in the formation of the planet. Scientists in the past have rejected this hypothesis because when earth formed, it was in a molten state. "Our planet probably experienced its hottest temperatures in its earliest days, when it was still colliding with other rocky debris (planetesimals) careening around the solar system. The heat from these collisions would have kept the Earth molten, with top-of-the-atmosphere temperatures upward of 3.600 degrees Fahrenheit."[50] Most scientific theories about the early history of the earth state that after earth initially formed, it was too hot to contain any water. "The earth formed under so much heat and pressure that it formed as a molten planet."[51] After rejecting the idea that water formed in the earth in its initial creation, scientists are now proposing same idea again, as they run out of other viable options. In this latest twist of which theory best explains where earth's water came from, scientists propose that water was trapped in the dust grains that comprised the dust cloud (disk accretion theory) that surrounded the newly formed sun. This dust water somehow survived the tremendous heat of our newly formed planet. Of course, the disk accretion theory of how planets formed has been rejected because the new data obtained from the Kepler telescope shows that the disk accretion theory will not fit other newly discovered solar systems that have multiple planets. Looking at the nuts and bolts of these three theories of where earth's water came from, none of them make any logical scientific sense. Perhaps in time, scientists will figure out where earth's water really came from.

Chapter 5

A SIMPLE CELL

Three point eight million years ago, on a warm cloudy day, one-half mile from the shore, on the tidal flats of what would one day be called First Encounter Beach in Cape Cod, two feet beneath the warm mud lay the primordial ingredients of the first bacterial cell. Suddenly, lightning strikes the mud ten feet away from this spot, sending one billion volts of electricity coursing through the sand and mud, producing the final catalyst needed to jump start the first bacterial cell into a living, breathing, self-replicating cell. What a miracle. Whether or not the first cell arose in Cape Cod or Russia does not matter, what matters is that according to the theory of evolution, somewhere in the world 3.8 billion years ago, a *simple cell* arose by time and chance through ordinary chemical processes. The theory of evolution hangs on this one small often under disused facet of the beginning of evolution. It is really the heart and soul of the theory of evolution. Could life really arise by time and chance, producing what evolutionist have continuously purported as a simple cell?

What does simple really mean?

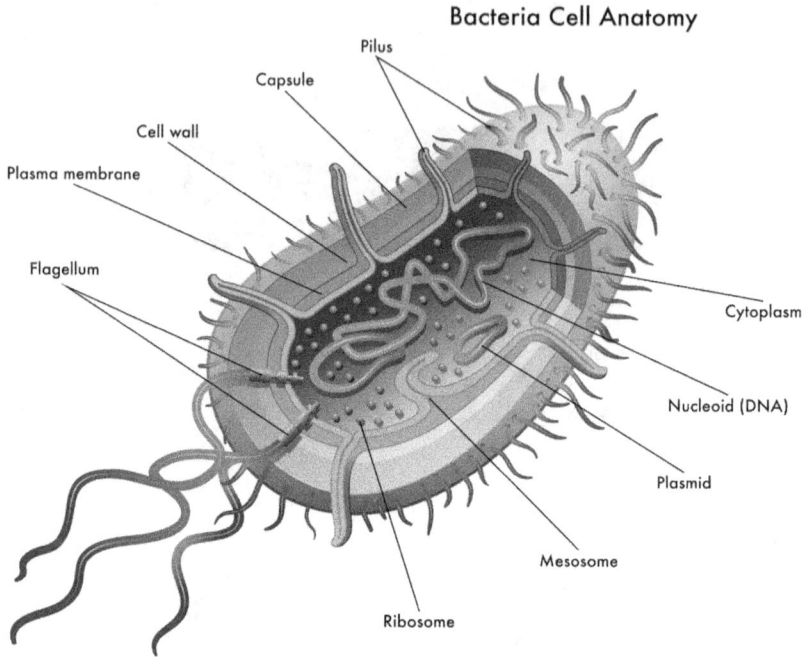

Bacteria Cell Anatomy

When I was in high school, I pictured a simple cell as a puzzle containing a dozen or so parts, comprised of an organic blob full of chemicals that given enough time, the primordial seas would jumble, toss, and eventually bring these chemicals into some type of order that gave rise to the first living cell. Theoretically, the idea of the first living cell being a simple cell has been very well ingrained in all people who have ever completed public education in America. After all, Stanley Miller almost created life in a test tube, "scientist say," or did he? What Stanley Miller did was try to recreate the primordial earth's environment, and he was able to create a number of amino acids through a very carefully controlled experiment. If you understand what constitutes the mechanics of a single-cell organism, creating a few amino acids is a very far cry from creating a living self-replicating cell. As I elaborate on what is truly required for chemicals to bond together to create the first living cell, I believe you will find the basic biological structure of the simplest living cell beyond amazing.

In the above image of a bacteria, I admit that on paper in two-dimensional form, a bacterium looks simple. Bacteria do not contain a nucleus and organelles that are bound by a membrane. Instead, bacteria are called prokaryotes and have a nucleoid which contains the cell's DNA and ribosomes within the cell's cytoplasm. And yet this simple-looking cell is composed of fifty billion atoms.[52] Requoting a statement I made earlier in this book, it would take just under thirty-two years to count to a billion at one-second intervals. A billion is not a number that our mind can really grasp, let alone fifty billion.

Let's look back at the beginning, according to the evolutionist hypothesis, that life began 3.8 billion years ago in a primordial sea. Imagine that you are walking along the ancient seashore of Cape Cod. There are no seagulls, no seals, no sandpipers, no grasses, no trees, no worms or seaweed, and virtually no oxygen (O_2). Scientists theorize that earth's early atmosphere contained carbon dioxide, water vapor, hydrogen, and nitrogen with no free oxygen. This would prevent any possibility of an ozone layer forming, which would protect the earth from the sun's harmful ultraviolet rays. It is in this type of environment that the first living cell had to arise by chance. Scientist believe that the first type of bacteria cells to evolve were cyanobacteria.

CYANOBACTERIA

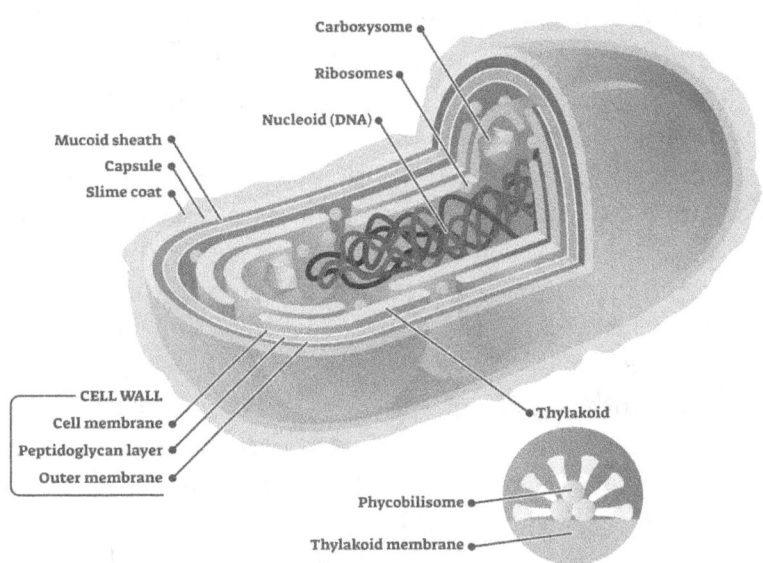

Through the process of photosynthesis, cyanobacteria take sunlight, carbon dioxide, water, and converting light energy into chemical energy that is stored as glucose—sugar. They also produce oxygen as a respiration byproduct. Scientists believe that cyanobacteria were the primary producers of the original oxygen in our ancient atmosphere. Whether the first living cell was a cyanobacteria or some other type of bacteria is not critical to the origin of the living cell. What is critical is, was it scientifically possible? Could life truly begin by random chemical reactions given enough time and chance. What constitutes life? Specifically, the first living cell, which most scientists agree, were bacteria.

This is a difficult question. Most scientists agree on the following stipulations of life for a living cell are the ability to grow, respond to stimuli, acquire food to produce energy, remove food waste, and reproduce itself. In the time of 1800s, when the idea of a simple cell was proposed by Charles Darwin in *On the Origin of Species*, scientists did not possess electron microscopes that now have the capability of magnifying up to ten million times. We can see what Darwin could never see, clear images of bacteria on an atomic level. Certainly, what we are able to see now is the fact that the smallest living cells in the world are not simple in any sense of the word.

As I have studied this subject of the first living cell, I have learned much. Microbiology was never a science that I truly liked, but I have found it to be far more interesting than I would have imagined. To start with, DNA is just an amazing part of every living cell beginning with what is supposed to be the simplest cell, the bacterium. When theorizing about what is needed to create the first living cell, I imagine all the parts that comprise a bacterium all have to randomly come together at one pinhead point at just the right time in just the right sequence. The heart and brain of every cell is its DNA. In the alleged simple bacteria, we need only two to three million base pairs of DNA. What is a base pair? They are the letters of all DNA code that arranged together in sequences that have meaning called genes. The building blocks of DNA are called nucleotides, ACGT. They stand for adenine, thymine, guanine, and cytosine. These nucleotides only pair with adenine to thymine and cytosine to guanine with a hydrogen bond. Whatever the heck does that mean? As these nucleotides bind together in base pairs, they form the twisted ladder of the DNA molecule. One strand of the DNA molecule is called a chromosome, which contains genes. Each gene codes for specific proteins. In humans, each of our cells contain twenty-three chromosomes with hundreds to thousands of genes per chromosome. Each gene contains the information a cell needs to build protein molecules, which form the structures of the bacterium and all living things. The sequence of the DNA base pairs becomes a language that tells a cell how to build a protein molecule. Each protein molecule is comprised of twenty different amino acids arranged in a very specific three-dimensional sequence hundreds of amino acids long. The sequence of DNA base pairs in each gene of a chromosome determines individual sequences of amino acids, which are necessary to build specific proteins. In a single bacterium, scientists estimate there are three hundred to five hundred different types of proteins. In a human, there are up to twenty thousand different types of proteins.

Understanding the information stored and encoded in DNA becomes so complicated even from the simplest cell in the world, the bacterium. Each bacterium has a single chromosome with five hundred to two thousand genes. Within each cell, there is a built-in computerlike process that splits the DNA molecule at the site of a gene, allowing another molecule RNA to make a copy of a gene, which is turned into messenger MRNA, which is sent to the ribosome to assemble the amino acids to create proteins. What happens in all living cells is just incomprehensible, even starting with what is considered the first living cell.

So how did the first living cell, a bacterium cell, assemble itself by chance? The DNA of a simple bacterium is a code of two to three million base pairs of adenine to thymine and guanine to cytosine arranged in a very specific code that will tell the bacterium how to make every part of itself and carry on the process of acquiring nutrients and expelling waste, building proteins, and reproducing and making exact copies of itself. It's the DNA that is the key. *It's a com-*

puter program. If a bacterium does not have a perfect DNA code, it will not live long.

The ultimate dilemma in creating the first living bacterium is that protein cells do not assemble themselves randomly. A protein molecule is a very complex molecule composed of twenty different amino acids carefully stitched together in the ribosome with very specific directions from the messenger RMA, which made an exact copy of a single gene encoded in a section of the DNA chromosome of the bacterium. What evolution requires is that all the components of the first living cell to have to randomly create each part and assemble themselves together at a pinhead point in the ocean or sea mud somewhere in the world 3.8 billion years ago.

In biology, most students have heard about the Miller experiment, where in eleven out of twenty amino acids were created in this experiment. This has been hailed as proof that life can almost be created in the laboratory. This is a very false statement. Eleven amino acids do not come close to creating a single-protein molecule let alone a DNA chromone with two to three million base pairs representing the code of life for a single bacterium. Mathematician Fred Hoyle have calculated the odds of a single-cell bacteria forming by chance is one in ten to the forty thousandth power.[53]

The number of all electrons in the entire universe is calculated at only ten to the eightieth power. The chance of just one protein forming by chance is one in ten to 10,130 power.[54]

If anyone researches the odds of just one protein forming by chance, they will find out that just one protein forming by chance is impossible. I mean it is absolutely completely impossible. Mathematically, it can never happen. A protein forming by chance in any experiment man has ever attempted has never happened nor can it ever happen unless that protein is assembled together inside the protein building center of ribosomes in the cell, following the DNA program built into each cell.

Taking all of this information, here is what is required to build the first living cell. In some random spot of ocean water or mud in the area the size of a pinhead, a DNA molecule with two to three million base pairs of code has to arise and assemble itself by chance.

This code is going to tell that bacteria how to assemble all its protein molecules; how to acquire food; how to communicate with other bacteria; how to build the cell wall, the cell membrane, the ribosome, the plasmids, the pili, and so much more; how to create energy it needs to live; how to get rid of waste; and ultimately, how to reproduce and exact copy of itself. Let's say this happened once, and now there is this perfect strand of DNA living out in the mud somewhere. Now the DNA needs to be surrounded by the cytoplasmic fluid containing a fully functional ribosome to build protein molecules and some plasmids all within that pinhead of space. Now the DNA, ribosomes, plasmids, and cytoplasmic fluid need to be surrounded by a cell membrane and then a cell wall, each composed of perhaps millions of protein molecules to build those two components of which a single-protein molecule is impossible to form. Then the cell wall needs to be surrounded by a capsule with pili attached. Most bacteria have a flagellum for locomotion. A single flagellum is so complex todays engineers could never have designed something so complex. So all of these actions I just described need to happen in sequence quickly in a spot the size of pinhead. If the components do not combine quickly into a cell, they will break down. Now even if all of this somehow miraculously happened, the first living cell would have to float up toward the surface of the ocean water to obtain solar energy, which would kill it within minutes because there is no ozone layer to block the ultraviolet rays of the sun.

A last tidbit of information. The law of biogenesis states that "life only arises from other living things by means of reproduction" (Louis Pasture). Pasture proved that only living things arise from other living things. The law of biogenesis states this. Life has never ever been observed to arise from nonliving matter. It is only for the theory of evolution that scientist have to make an exception and say that life arose by chance once. Evolutionist call this theory *abiogenesis*. Life can arise from nonliving matter. Choose for yourself which theory you want to believe. The laws of probability clearly predict that for all the individual components of the first living cell to arise by time and chance, it is absolutely impossible. The DNA molecules that are packed into the chromosomes of all living things represent a code, a preprogramed computer program that tells a cell how to do everything it does. In the simplest of all living cells, the bacterium, this program has two to three million base pairs of nucleotides adenine to thymine and cytosine to guanine arranged in a language that tells a cell everything it needs to do. Evolutionist call this luck; I call it the fingerprint of God.

Chapter 6

WHAT'S NEXT?

In college, I used to believe that the basic tenants of the theory of evolution were more or less true with one exception. I believed that God created the first living cell and guided the process of evolution one step at a time for billions of years. Scientifically though, such an idea just does not work. As I have spent almost forty years studying and researching this subject, I find that the biblical description that God created everything in seven days is the one reality that makes the most logical scientific sense. You may say the Genesis account is crazy and the stuff of fairy tales, but I say not so. I say the big bang, the formation of the planets, the acquisition of water on our planet, and the formation of the first living cell by chance is a lot more fictitious. Scientifically, none of those sub-theories of evolution pertaining to the origin of the universe and organic life have been proven by science and the scientific method. An example is the law of biogenesis and the theory of abiogenesis. One theory, the law of biogenesis, states that life cannot spontaneously generate from nonliving matter, while the other theory, abiogenesis, states that life can spontaneously generate from nonliving matter, given the right primordial conditions and time. If the impossible had occurred, and abiogenesis had occurred in a primordial earth with no oxygen, no ozone layer would have existed. No ozone is no life on planet earth. Yet life began on planet earth, evolutionists say, with cyanobacteria that obtain their energy through photosynthesis-producing oxygen as their respiration product. Scientifically though, how could cyanobacteria have ever survived and multiplied near the surface of the oceans without a layer of ozone surrounding the planet? It would have been impossible. These ultraviolet rays from the sun would have killed all cyanobacteria that were near the surface of the ocean water long before they could have multiplied and created an ozone layer. Again, the step-by-step processes the theory of evolution requires just does not work, nor is it scientifically logical and feasible. Creationism is based on what is written in the Bible. Either God is a myth and the entire Bible is the imagination of men's heart, or God is real, and he created the heavens and the earth and all life on earth. According to chapter 1 in the Bible, God created the universe, the water on planet earth, the air, the plants, the birds, and all living creatures in seven twenty-four-hour days. This is what creationism means. It means God created everything. Within the Christian community, there is much disagreement about the theory of evolution and the age of the earth. Did God just create the first bacteria and let things go? I have come to think of the creationism worldview in this way—if God was smart enough to create DNA, I certainly think he would be smart enough to write one book to mankind and write that book without any errors. Hence, I believe the Genesis account that God created everything in seven literal twenty-four-hour days is true.

It is the DNA that is the key to all living things. "Genes are composed of DNA, and it is predicted that there are over three billion base pairs of DNA in the human genome. A human body has approximately thirty-seven trillion cells, so if you were to line all of the DNA found in every cell of a human body, it would stretch from the earth to the sun one hundred times!"[55] This is a quote from the national human genome project. The amount of information stored in just one cell of the human body is beyond our comprehension.

But if you took all the blood vessels out of an average child and laid them out in one line, the line would stretch over sixty thousand miles. An adult's would be closer to one hundred thousand miles long (Franklin Institute). Your heart beats one hundred thousand times a day, thirty-five million times a year.

During an average lifetime, the human heart will beat more than three billion times. The adult heart pumps five quarts of blood each minute throughout the body, adding up to approximately two thousand gallons of blood each day. During your lifetime, it will pump about eight hundred million pints of blood or about one million barrels. That's enough to fill more than three super tankers (Arkansas Heart Hospital).

Each facet of the human body is an engineering masterpiece. For example, "on average, each of the one hundred billion neurons in your head has about five thousand connections with other neurons, creating a huge network of about five hundred trillion synapses. Like a computer network built from five hundred trillion transistors, each representing a 'bit' of information depending on whether it is 'on' or 'off'" (Rick Hanson, PhD, 2007 www.WiseBrain.org).

A piece of brain tissue the size of a grain of sand contains one hundred thousand neurons and one billion synapses, all communicating with one another (Brain MD life, April 20, 2016).

At the time of this writing, the fastest supercomputer in the world is the *Tianhe-2* in Guangzhou, China, and has a maximum processing speed of 54.902 peta FLOPS. A peta FLOP is a quadrillion (one thousand trillion) floating point calculations per second. That's a huge amount of calculations, and yet that doesn't even come close to the processing speed of the human brain.

In contrast, our miraculous brains operate on the next order higher. Although it is impossible to precisely calculate, it is postulated that the human brain operates at one extra FLOP, which is equivalent to a billion, billion calculations per second (Science ABC).

Second in complexity to the human brain is our eyes. Most current digital cameras have a five to twenty megapixels, which is often cited as falling far short of our own visual system. This is based on the fact that at twenty-twenty vision, the human eye is able to resolve the equivalent of a fifty-two-megapixel camera, assuming a sixty-degree angle of view (Cambridge in color 2005–2019).

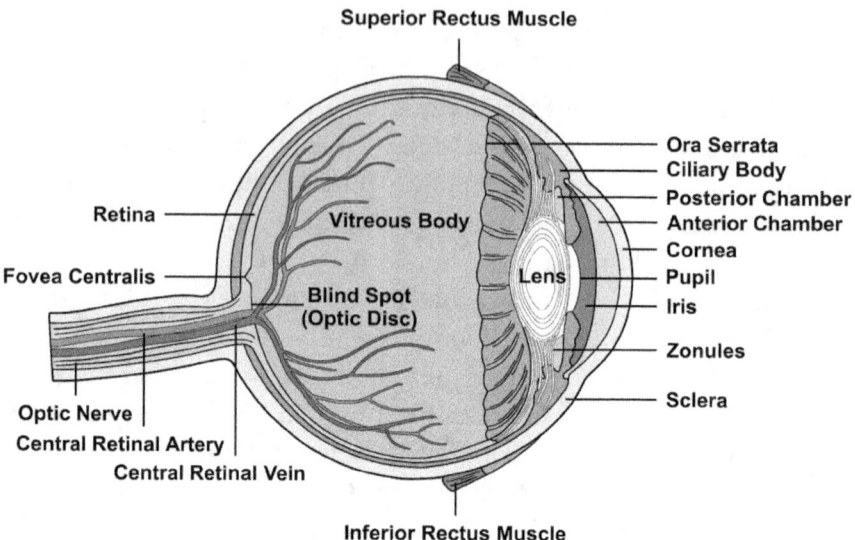

How eyes function is either an absolute miracle of evolution or design by an intelligent creator—God. Light enters our eyes throgh the cornea and passes through our pupil, which opens and closes like a camera to allow more or less light through the pupil, depending on how bright the light source is. This light is then focused by a lens that expands or contracts depending upon whether we are looking at near or far objects. Light then continues throught the vitreaous chamber and is focused on the the wall of the retina, which is composed of millions of light sensative cells called rods and cones. These light sensitive cells turn the light into electrical impulses and send these implulses throurgh millions of optic nerve cells in the optic nerve itself to the brain, which turns these electrical impluses into an image.

How does all this really work? We can figure out how a computor can take electrical implulses and turn them into pictures, but how does our brain, which weighs just a few pounds and looks like a bowl of macaroni and cheese, change light images into pictures and pictures into memories, memories that we can recall fifty years later? How do our brains really store memories and recall memories? Going back in the alleged time chamber of evoulution 3.8 billion years ago, the first one-cell oraganism bacteria evolved into a thirty-seven-trillion-celled human being in just a few billion years.

How did evolution do it? Charles Darwin in *On the Orgin of Species* coined the idea of survival of the fittest. Today's evolutionist have refined that idea to natural selection where organisms that are better adapted to their environment will survive. Given enough time, millions of years, simple organisms are supposed to evolve into more complex organisms. I have always wondered what the first bacteria evolved into. Of all the living organisms on our planet, bacteria are the most prolific. They can reproduce in twelve minutes to twenty-four hours with the average life span of approximately twelve hours. The scientists from the University of Georgia estimate the number of bacteria on our planet to be five million trillion, trillion. That's a five with *thirty* zeroes after it. There are far more bacteria on earth than there are stars in the universe.

Therefore, with five to the thirtieth power number of bacteria on earth and a reproduction and death rate of twelve hours, that will yield five to the thirtieth power number of new bacteria every day. That is more stars in the universe number of chances for a bacterium to evolve into something else every single day. *And yet they don't.* If a single bacterium in the last one hundred years had been known to change into a new species, it would have been the world's greatest news for evolutionist.

Thirty years ago, Michigan State University Researcher Richard Lenski added his now-famous bacteria to twelve inaugural flasks, a process he and his team of lab technicians and students have been repeating daily ever since. Through the years—and more than sixty-eight thousands of generations of bacteria—the bacteria had evolved to eat something other than glucose, standard fare for all the other generations of their bacteria. The experiment also has earned the MSU Hannah Distinguished Professor of Microbiology

These special bacteria had begun consuming a food source called citrate. In the halls of evolution, this was hailed as absolute proof that evolution occurs. "I say so what!" The bottom line is they are still bacteria.

What evolution requires for the creation of each and every new species of plants and animals is the creation of a brand-new DNA code for every single living organism, again including both plant and animals. The simplest living thing in the beginning of the world, bacteria, had a computer code (DNA) with two to three million lines of code that is far more complicated than any man could have ever imagined. Now that simple one heck of a complicated cell has to change into an organism with millions of cells. How on earth did it do this? How did a single cell evolve into a living organism with a brain, heart, eyes, circulation system, blood, digestive tract, and all that we are? It is a major leap from a single cell.

Within the halls of evolution, it seems to me that the theory of evolution just skips the complicated parts that are truly unexplainable and scientifically implausible. That would include how the first cell arose and how the first multicellular organism evolved from a single cell organism and how all other life-forms in the entire planet evolved. Here is a prime example. The evolution of the whale.

A prime example is the evolution of the whale. In most scientific books about evolution, they will include an artistic drawing of a land-based cow or something similar, that a series of pictures will add fins, flippers and eventually a tale. This is not based on fact, but imagination. Fossils do not support these artistic drawings. If you go to YouTube, you can watch a video of some unknown and weird-looking land-based mythological mammal that somehow started to evolve webbed feet. That's not going to work to well while living on dry land. I would have thought the first one to have evolved would have been eaten pretty darn quick. Millions of years later, these flippers evolved into fins and a fluke. In a whale, the tale has two parts called flukes. It was obviously quite a fluke that these flukes ever evolved. The word *fluke* also means that something happened quite accidentally. This is what evolutionist call *science*. All life-forms are a colossal set of mutations that created entire new sets of DNA like the evolution of the whale. All I can say is OMG! Our life is a fluke. Jumping back to the second form of life on earth after a simple bacterium, I have become quite perplexed trying to figure out what the second life-form on planet earth was.

As I have researched this idea of what was the first multicellular organism on the World Wide Web, I found virtually nothing. I found examples where groups of single cells form colonies that work together, and therefore, this somehow led to the first multicellular organism. Still, what was it? As I look up the evolutionary tree of life, I cannot make any sense out of what they show in these weird tree of life diagrams. These diagrams start at the bottom with a single-cell organism and have branches going everywhere with one branch turning into plants and the other branches turning into animals. The diagrams are usually quite vague with a lot of missing blanks. I just want to know what the next living organism that evolved after the first single cell. Was it a simple worm, a snail, a clam, a jellyfish? As I read about the beginning of evolution, the theory states that life began about 3.8 billion years ago as a simple cell. This idea is fairly consistent, and another 1.7 billion years later more or less, multicellular organism arose, but what those multicellular organisms were is a mystery.

A fossilized life-form that existed 558 million years ago has been identified as the oldest known animal, according to new research. The new fossils, of the genus *Dickinsonia*, are the remains of an oval-shaped life-form and part of an ancient and enigmatic group of organisms called *Ediacarans* (The Guardian 2018).

Whatever that fossil is, it is probably composed of millions, if not billions, of cells with some type of circulation system, brain, digestive tract, and probably more organs. Why would a single-bacterium cell suddenly change into such an organism? There should be preceding organisms but as scientist are claiming, this is perhaps the oldest known animal. What changes in the DNA of a bacterium would need to occur to suddenly take a single-cell bacterium and change it into an animal with millions of cells all working together. Any way you think about this problem, it is scientifically impossible. There is no logical thinking that a one-celled organism can one day accidentally change into a million- or billion-cell organism. This is just another mystery that evolutionist have not yet solved.

Chapter 7

EVOLUTION

A *quick review*. In the beginning, thirteen billion years ago, there was nothing, and out of nothing, a singularity formed, containing all the matter of the universe. It was infinitely dense and infinitely hot and as small as a pinhead. The terms *infinitely hot* and *dense* are not scientific at all. For matter to be created out of nothing is not scientific at all. The first law of thermodynamics states that energy/matter cannot be created or destroyed. This creates a problem right from the start for the big bang theory. All scientific explanations I have heard about the big bang singularity was that the gravity generated from it was so intense that not even light could escape. Why did this thing blow? I remember reading a scientific American magazine one day at my barber shop, and this question of why the big bang exploded was being discussed in an article. One theory proposed about the initial explosion was that there was some antigravity matter present for a moment, then voila, it exploded. That's how it all started.

Then about four to five billion years ago, the solar system formed with all the planets. The best explanation for the formation of our sun and the solar system planets was the disk accretion theory, which postulates that the spinning dust in space spread out like a record, eventually condensed into the sun and the planets of our solar system. They all had to spin in roughly a plane and rotate in the same direction, which is fine for our solar system, but astronomers have discovered with the aid of the Kepler telescope, thousands of planets in other solar systems do not all rotate in a plane or in the same direction. Some planets rotate on the opposite to the direction of their sun or in an elliptical path, not at all in a plane as our planets are. This new information has destroyed any logical planet formation theory. I find this disheartening.

Then once our planet formed about 4.5 billion years ago, it is believed to have been a molten mass that initially had no water. Where did the earth get its water? Asteroids! All 1,665 million cubic miles of them. Earth has an estimated 333,000,000 cubic miles of water according to the US Geological survey. Asteroids have an approximate 20 percent water content; therefore, to get one cubic mile of water, we need five cubic miles of asteroids. Hence, the 1,665 million number of cubic miles of asteroids are needed to get the 333,000,000 cubic miles of water. If this theory is true, where did the four-fifth of asteroid waste go? To get that amount of water, we would generate four oceans full of asteroid waste. Again, there is not a truly logical or scientific reason why we have water on this planet.

Geologists unequivocally agree that the age of the earth is 4.5 or 4.75 billion years old. A one-mile-thick sedimentary layer of rock covers most of our planet, and the layers found in the Grand Canyon are said to represent approximately 1.8 billion years of our planet's history. The present is the key to the past is the hallmark statement in geology. This statement means the sedimentation rates we see in the world today should have been the same for most of earth's history. Therefore, the average sedimentation rates observed today on continental shelves can be estimated at one-thousand-year increments and should equal a deposition rate that will yield a 1.8 billion years. Extrapolating this sedimentation rate should mathematically add up to a thickness of one mile of sediment, thus equaling 1.8 billion years of earth's history. Using simple math, the sedimentation rates on the continental shelf today average a deposition rate of thirty centimeters per one thousand years. Dividing this rate into a mile of sediments equals an age of 5.4 million years, not 1.8 billion years. If you add a global flood where most of these sediments were deposited in a year, you would have a very young earth. Certainly, 5.4 million years is nowhere close to 1,800 million years or 1.8 billion years. I remember in geology class in college how we were taught that it took oil and gas millions of years to form, yet natural gas forms in landfills on a daily basis. In living animals, natural gas can form in hours by bacterial action during food digestion. Teenage boys have found this gas to be flammable. Another conundrum is the fact that scientists are discovering soft tissue in many, many dinosaur bones, indicating that dinosaurs are not possibly millions of years old.

Then 3.8 billion years ago, the first simple but unbelievably complex one-celled organism formed by time and chance, a cyanobacteria. This type of bacteria can live without oxygen, and it gives off oxygen as a respiration product. Once it was born and floated to the ocean surface, it would have died without an ozone layer surrounding earth. Where did the ozone layer that surrounds our planet come from? It came from cyanobacteria. Think about that. Louis Pasture discovered the law of biogenesis which states, "Life does not ever arise from nonlife." That law has never ever been violated, except with the theory of evolution. It is the birth of the first living cell that no evolutionary theory can explain. It is mathematically and scientifically impossible. The DNA code of all living things could never have arisen by time and chance. The so-called simple cell (a bacterium) is composed of two to three million base pairs or lines of a computer code. This DNA tells the cell how to do everything. It tells the cell how to make every molecule that comprises the next cell that will be an exact copy of itself. These copies are so perfect that as long as scientist have been studying bacteria, they have never evolved into a different species. Every day the simple bacteria have 1,000,000,000,000,000,000,000,000 chances to evolve into something else, yet they don't. Virtually every living organism that has billions of cells in its anatomy will never have this many chance to evolve, and yet we still have the same simple but impossibly complex bacteria living today. It has not evolved into a new species.

I remember reading a question some student asked online. "Is evolution still occurring today?" The answer by the professor was, "Yes, we just don't see it happening because it takes about two million years for a living thing to evolve. Every single day, the bacterium will have more chances to evolve into something else than all living organisms, yet they don't."

I continued to ponder that question, and I thought to myself, *Okay, say what that professor said about evolution taking two million years to happen for any living thing is true. Therefore, we cannot see it happening. Therefore, all living things are on a two-million-year-old evolutionary time clock, meaning every two million years, most living things would evolve into something else. How would it be possible for all living things to be on the same two-million-year-old time clock?*

BBC news, science, and environment on August 23, 2011, states, "The natural world contains about 8.7 million species, according to a new estimate described by scientists as the most accurate ever." With that many species, there should be millions of species on a different time clocks with thousands of species somewhere in the middle of evolving. These would be the missing link species. There should be some land animals that are now living in the water with webbed feet and its tail turning into a flipper. I heard of this idea was called a sea cow at one time. Well, let's be honest, we do not see species in the middle of evolving today. The idea that it takes too long to observe does not hold water. There should be thousands of species alive today in the middle of evolving, but there are none.

Then there is the fossil record. If evolution is true, there should be billions of intermediate forms of living things in the process of evolving into something else. Again, these are the missing link fossils of which there are not many. To my knowledge, there are a dozen so-called missing link fossils such as archaeopteryx, the missing link in bird evolution, yet all these missing link / transitional fossils are all disputed. Some paleontologist will say that there are no any intermediate forms. One of the most famous quotes by a paleontologist was by Dr. Colin Patterson, who was the senior paleontologist at the British Museum of Natural History. He stated "that he could not find a single transition fossil that he could make a watertight argument as being a transition fossil" (Sunderland, L., *Darwin's Enigma*, Master Books, Arkansas, USA, pp. 101–102, 1998). Patterson's letter was written in 1979. This is absolute truth. The fossil evidence for evolution is simply not there. The real verifiable evidence for the theory of evolution is transition fossils. There should be trillions of them or at least billions of transition fossil, but no there is only a handful. Of the handful of transition fossils, half of those are turning out to be fraud and remaining transition fossils are questionable.

In the famous Scopes trial in 1925, a substitute high school teacher, Thomas Scopes, was accused of teaching human evolution in a state-run school. The resulting trial known as the Scopes trial presented evidence for human evolution, including Piltdown man and Nebraska man. The evidence for Nebraska man was a human-looking tooth that was supposed to be a fifty-thousand-year-old ancient man.

From a single tooth, scientists were able to draw a picture of what Nebraska man looked like. You can see pictures of Nebraska man by simply looking up images of Nebraska man on the internet. This is science fiction drawing a missing link man from a single tooth. This tooth was used to create Piltdown man used in the Scopes trial against the state of Tennessee fighting to allow evolution to be taught in public schools. They won. However, a few years later, it was discovered to be outright fraud. These are facts. Check them for yourself.

Another critical fact about evolution is how did species change from one species to another? Mutations. In the beginning, evolution had to find a way to create the perfect genetic computer program built into all living things called DNA. It is the DNA of all living things that directs the creation and reproduction of all living things. Creationist claim that God is the inventor of the DNA in all living things, while evolutionist believe that time and chance mixed with beneficial mutations is the scientific reason for why all living things are. Beginning with the first simple cell, the bacteria, knowledge has proven the absolute impossibility of the first living cell beginning by time and chance. It is also impossible to see any clear path to the creation of the first living multicellular organism. How was it possible for the simplest of multicellular organisms to develop the first heart, the first eyes, the first blood cell, the first mouth and digestive system, the first brain? None of these fundamental parts of the anatomy of all living things can be explained by time and chance accumulations of beneficial mutations. Each part of the basic organs of all living things is written in a computer program of millions of lines of DNA code that is so utterly complex that no man could ever have conceived of such ideas with all of mankind's knowledge today.

The key to evolutionary change that I was taught in school was natural selection, where organisms better adapted to their environment were able to survive and pass those more adaptive traits on to their offspring. Given enough time, a new species would eventually arise. The peppered moth was the greatest example of evolution ever shown while I was in school. During the industrial evolution, when pollution darkened trees that the normally light-colored peppered moths lived on, caused the light-colored moths to be easily spotted by prey, and soon all light-colored moths disappeared as they were not environmentally advantageous anymore. However, the dark-peppered moths were now environmentally favorable, and soon the genome of dark-colored moths became the primary population of peppered moths. When the pollution of the industrial revolution had improved, the population of peppered moths shifted back to light-colored moths. For many years, this example of adaptation of a species has been touted as proof of evolution. Evolution takes this example and extrapolates the results millions of years into the future until a new species of moth arises, a dark-colored peppered moth. I don't see this as producing a viable new species. If it does produce a new species, it would take millions of years, and the fossil record would contain billions of intermediate fossil types, which just are not there.

It takes much more than some pollution to create a new species. It takes completely new lines of genetic DNA codes. This is not just a few point errors in the strands of DNA. To create new different species, thousands to millions of lines of new DNA code are needed. Without a designer, i.e. God, we are left with random changes in the DNA code that have to be cumulatively beneficial. This would involve again millions of minute changes to the DNA that are cumulative to both male and female organisms at the same time so that these changes can be passed on to their offspring. In the realm of mutations, it is commonly known that detrimental mutations outnumber beneficial mutations thousands to one. In addition, known examples of beneficial mutations, there are only a handful of examples. In humans, mutations help improved resistance to malaria and HIV, increased bone density and resistance to heart disease. These are nice mutations but hardly anything tangible to help create a new species. Again, research beneficial mutations in animals, the examples of beneficial mutations are quite elusive to nonexistent.

Family trees in evolution. They used to have family trees that would connect the so-called dots. If we truly evolved from a single cell, there has to be a line of animals to connect the dots.

Bacteria, sponge, jellyfish, fish, lizard, rat, cat, marsupial, chimp, ape, human. That's the best I can come up with. Online, I find nothing as detailed as I just proposed. If life began in the ocean, we must have turned into a fish that eventually crawled onto land. To evolve from a one-cell animal to a human, evolution had to bridge four major class of animals. A one-cell animal turns into a multicell animal, then a worm or jellyfish, then into a fish. That fish crawls onto land and is now an amphibian, then the amphibian turns into a reptile—lizard. The lizard is a cold-blooded scale-covered animal with a three-chambered heart that evolves into a mammal that is warm-blooded and has a four-chambered heart, possibly a rat. That rat evolves into a catlike animal that changes into some type of marsupial, which evolves into a monkey, ape, and then humans. This is our heritage as an evolutionist. When we die, we become dust? We have no soul. Sin is a man-made idea. There is no heaven or hell. Our morals are what we decide as a society. Everything is okay until you die or your wife dies or your best friend dies because then you are no more. Nothing, just a rotting carcass six feet underground.

Oh, people, can you not see that you have been duped, conned, and lied to? Evolution is man's attempt to run away from the God who created you and I with a divine purpose. We are created to live forever with God in paradise one day. Evolution is just a theory designed to take away your heart and soul and keep you in the dark about having a life filled to the full walking and knowing God, your creator.

I could write pages and pages of facts disputing evolution. Evolution has answers to nothing. The idea that a dinosaurs' scales turn into feathers, a three-chambered heart turns into a four-chambered heart, and the dinosaur sprouts wings, a lighter skeleton, infinitely better eyes, and one day is a bird, then you have been deceived. There is only about a handful of fossils that are allegedly transition fossils showing some kind of transition of species, and those fossils are all challenged as to whether they are truly transition fossils. It is as Dr. Patterson, a senior paleontologist at the British Museum of Natural History, said that there is not one conclusive fossil that he could claim was a true transition fossil. Evolutionist can look for those fossils forever. If evolution is was really true, transition fossils would be found by the billions, and here we are today with a handful of questionable transition fossils. And from a pig's tooth Nebraska man, evolution won the write at the Scopes trail to be brought into our public schools to teach generations an alternative to the God who created us.

Science does not have the answer to life. It cannot answer where all of the matter of the universe came from to make an alleged big bang. Science does not know why the big bang blew up. Science does not know how planets formed with the latest planetary information we now have. Science does not have the answer of how the earth got our moon. Science does not have the answer of how and where earth got its water. Science does not have a clue how the first cell arose with such an amazing DNA program built into it. Lastly, science evolution has no clue and no fossil evidence to support how a one-celled bacteria evolved in a human in 3.7 billion years.

Choose life, not evolution! In God, there is life.

Notes

1
2
3
4
5
6
7
8
9
10
11
12
13
14
15
16
17
18
19
20
21
22
23
24
25
26
27
28
29
30
31
32
33
34

35
36
37
38
39
40 National Ocean Service Http://oceanservice.noaa.gov/facts/oceanwater.html Revised March 22, 2016.
41 Dictionary.com (http://www.dictionary.com/) Chemical evolution.
42 Universe Today (http://www.universetoday.com/76509/how-was-the-earth-formed/ Dec. 23, 2015.
43 National Geographic news.com Http://news.nationalgeographic.com/news/2011/02/110222-planets-formation-theory-busted-earth-science-space.
44 Ibid
45 Ibid.
46 Same as 47.
47 Science News http://www.sciencenews.org/article/how-did-earth-get-its-water By Christopher Crocket May 6, 2015.
48 Marcus Woo Science 12/10/14 Https://www.wired.com/2014/12/rosetta-mission-comet-water-earth/.
49 Charles Q. Choi Space.com 2/2/13 Https://www.space.com/19681-dinosaur-killer-asteroid-chicxulub-crater.html.
50 Michon Scott, Rebecca Lindsey August 12, 2014 https://www.climate.gov/news-features/climate-qa/whats-hottest-earths-earths-ever-been.
51
52 Below is the article that appeared in the November 12, 1981 issue of Nature, page 105 (volume 294, number 5837). It is a text box in the "News" section.
53
54 Laurence D Smart BSc, Agr, DipEd, Grad, DipEd, Email: laurence@unmaskingevolution.com, Webpage: www.unmaskingevolution.com.
55

www.ingramcontent.com/pod-product-compliance
Lightning Source LLC
Chambersburg PA
CBHW072025230526
45466CB00019B/675